FOURTH EDITION

INTRODUCTION TO
STATISTICS

AMY B. MADDOX

Cover image © Shutterstock, Inc.

Kendall Hunt
publishing company

www.kendallhunt.com
Send all inquiries to:
4050 Westmark Drive
Dubuque, IA 52004-1840

Copyright © 2013, 2014, 2017, 2021 by Amy B. Maddox

ISBN 978-1-7924-6755-4

Kendall Hunt Publishing Company has the exclusive rights to reproduce this work,
to prepare derivative works from this work, to publicly distribute this work,
to publicly perform this work and to publicly display this work.

All rights reserved. No part of this publication may be reproduced,
stored in a retrieval system, or transmitted, in any form or by any
means, electronic, mechanical, photocopying, recording, or otherwise,
without the prior written permission of the copyright owner.

Published in the United States of America

CONTENTS

Chapter 1: Introduction to Statistics 1

Chapter 2: Discrete Probability Distributions 43

Chapter 3: Continuous Probability Distributions 67

Chapter 4: Sampling Distributions 89

Chapter 5: Estimation and Confidence Intervals for One Parameter 107

Chapter 6: Hypothesis Testing for One Parameter 133

Chapter 7: Hypothesis Testing for Two Parameters 165

Chapter 8: Categorical Analysis 189

Chapter 9: Simple Linear Regression. 209

Appendix: Answers to the Checking Your Understanding Problems 243

INTRODUCTION TO STATISTICS

➡ Section 1: Introduction:

What exactly is statistics? *Statistics is the art of collecting, organizing, summarizing, and describing data as well as drawing inferences from the data.*

- The goal of statistics is to draw conclusions about the real world based on observations one makes. In the language of statistics, the goal is to draw conclusions about a population based on what is observed from a sample taken from that population. This might sound simple, but due to the fact that many times, there is more than one phenomenon occurring, this can be quite a challenge.

- Consider the following example courtesy of Dr. John Seaman. Suppose you are concerned about the difficulties some couples have in conceiving a child.

  - It is thought that women exposed to a particular toxin in their workplace have greater difficulty becoming pregnant compared with women who are not exposed to the toxin. You conduct a study of such women recording the time it takes to conceive.

  - Of course, there is natural variability in the time it takes to achieve a pregnancy, which is attributable to many causes aside from the toxin. Nevertheless, suppose you finally determine that those females with the greatest exposure to the toxin had the most difficulty getting pregnant.

  - But, what if there is a variable you did not consider that could be the cause? No study can consider every possibility.

  - For example, it turns out that women who smoke while they are pregnant reduce the chance their daughters will be able to conceive because the toxins involved in smoking effect the eggs in the female fetus!

  - If you didn't record whether or not the females had mothers who smoked when they were pregnant, you may draw the wrong conclusion about the industrial toxin.

  - There is always natural variability in the data.

  - You must be careful with **extraneous factors**, which is anything that can affect the outcome, other than the treatment.

  - No study can consider every possibility. But you do want to control for extraneous factors!

- In statistics, the conclusions we make always come with some amount of uncertainty. i.e., we never really know if we made the correct decision. This is because we are dealing with a sample and not the entire population, i.e., we are not able to measure every woman's time to conception, only a sample of women.

- In statistics we quantify our uncertainty about our decision using probability, which we will discuss later. This allows us as statisticians to determine how "good" our decisions are.

Consider the following illustration: We have a specific population that we would like to know something about, so we randomly sample from that population, look at a picture of the sample, and then, with the idea of repeated sampling, create a population model that we use to draw conclusions about the population.

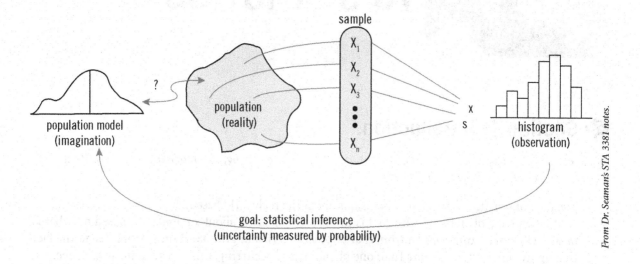

We can divide the study of statistics into two broad sections, descriptive and inferential. In descriptive statistics one is simply describing what they observed from the sample, whereas in inferential statistics, one is using the information discovered in the sample to draw conclusions about the population.

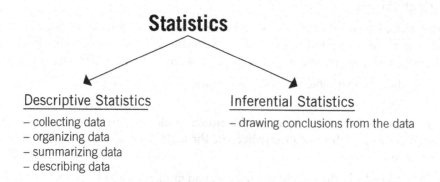

We will begin discussing the areas of descriptive statistics, but will spend the majority of the class in inferential statistics. I like to think of the notes in the beginning of the semester as giving us the "tools" we need to do inferential statistics. Obviously before you can begin drawing conclusions, you have to be able to collect and organize the data and understand the data. But first, we must learn some basic definitions used in statistics.

Definitions

- *Population* – the complete set of units we are interested in studying. The population may be people, animals, or objects. Most people tend to think of a population as containing people, but if I am interested in studying tires, then my population is ALL tires. (sometimes referred to as *Target Population*.)

    It can sometimes be a challenge to determine the population. Specifying the population is the first and one of the most important steps of statistics.

The size of the population is denoted by the symbol, N.

(Statistics is a language that uses symbols and notations to convey meaning. It is very important that you learn the "language" of statistics by learning the symbols and the correct notation.)

Example: What is the population in a political poll?

A. All eligible voters

B. All registered voters

C. All persons who voted in the last election

Answer: The answer is B. If we want to know who is going to win, our population is only those registered to vote because they are the only ones that can determine the outcome.

- *Sample* – a subset of the population.
- *Well Representative Sample* – a sample in which the characteristics of the sample match the characteristics of the population. (Our goal is to always have a well-represented sample, but as we will see, that is sometimes a challenging goal.)

The sample size is denoted by the symbol, n.

- *Parameter* – a numerical value based on the population. Unless we have information from every element of the population, the parameter is unknown.
- *Statistic* – a numerical value based on the sample. Using the information found in the sample, we can calculate the statistic.

In drawing conclusions, we use the statistic to draw conclusions about the parameter. We will learn later how to do this, but this is a very important concept to understand!

- *Variable* – a characteristic that takes on different values for different people, places, or things is labeled a variable. For a value to be a variable, it must potentially vary for all elements of the population.
- *Constant* – values in the population that do not vary are known as constants.

  - Is a parameter a variable or a constant?

    Answer:

  - Is a statistic a variable or a constant?

    Answer:

- *Probability* – measures how likely it is for something to occur. We can only calculate probabilities about random variables, not constants.

  - Can we calculate probability about a parameter?

    Answer:

Introduction to Statistics **3**

- Can we calculate probability about a statistic?
  Answer:

- *Random Variable* – A variable is a *random variable* if the exact value of the variable cannot be predicted in advance. This occurs when the value of a variable is obtained as a result of a chance factor.
- There are two "types" of variables:
  - *Quantitative Random Variable* – a variable that can be measured in the usual sense.
  - *Qualitative Random Variable* – a characteristic that cannot be measured in the usual sense; it can only be put into categories. (Also called a *Categorical variable*.)
- There are two types of quantitative random variables:
  - *Discrete Random Variable* – when the variable can only assume specific values.
  - *Continuous Random Variable* – when the variable can theoretically assume any value on a given interval.

  Example: Suppose my area of research concerns feet, specifically the size of my students' feet. What are two questions/things that I can ask or do to determine the size of your feet?

  Answer: 1.

  Is _____ discrete or continuous?

  Answer: 2.

  Is _____ discrete or continuous?

- Here is a flow chart showing the different types of random variables: It is important to understand the different types of random variables. The type of random variable will determine the method used to analyze the data.

4 Introduction to Statistics

Measurement and Measurement Scales

- *Measurement* – the assignment of numbers to objects or events according to a set of rules. There are four measurement scales that result from the fact that measurements may be carried out using different rules.
  - *The Nominal Scale* – "naming observations" or classifying them into various mutually exclusive categories, for example, male/female.
  - *The Ordinal Scale* – when the measurements can be ranked according to some criterion, for example, intelligence– low/average/high (Note that arithmetic operations are not meaningful.)
  - *The Interval Scale* – when the measurements cannot only be ranked, but the distance between any two measurements is known. There is no true zero point. For example, temperature using Celsius or Fahrenheit scale. 60°F is 30°F greater than 30°F, but 60°F is not twice as hot as 30°F.
  - *The Ratio Scale* – measurements in which equality of ratios as well as equality of intervals is known. (There is a true zero point.) For example, height, weight, length, time to complete a task. Ten minutes is twice as long as five minutes.

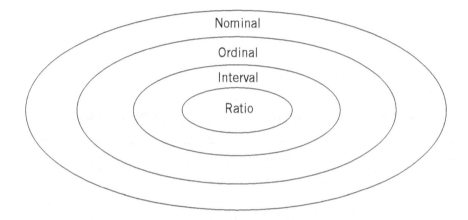

Checking your Understanding: (Answers are found in the Appendix)

Suppose Dr. Veen, a well-known oncologist, is interested in determining the average age of women when first diagnosed with breast cancer. Dr. Veen randomly samples 100 women that have been diagnosed with breast cancer and calculates their average age.

   a. What is the parameter?

   b. What is the statistic?

Section 1: Homework:

1. What is an extraneous factor?
2. Suppose Dr. Veen, a well-known oncologist, is interested in determining the average age of women when first diagnosed with breast cancer. Dr. Veen randomly samples 100 women that have been diagnosed with breast cancer and calculates their average age.

   a. What is the population?

   b. What is the sample?

3. For the following scenarios, determine if the random variable is Qualitative or Quantitative. If the random variable is Quantitative, determine if it is discrete or continuous.

  a. Dr. Vernon is measuring the height of 25 randomly selected kindergarten students.

  b. Dr. Johns gave a survey where he asked the subjects on a five point Likert Scale their opinion on the latest healthcare requirements.

  c. Dr. Ban is counting the number of kindergarten students that attended a certified pre-K program.

➡ Section 2: Descriptive Statistics – Collecting Data:

This is a very important area of statistics. If you have bad data, then no matter how great your techniques are, your results are going to be meaningless. We will briefly discuss some general techniques for collecting data, but we will by no means cover all the different ways you can collect data. This is just a small overview of the topic.

- Methods for Obtaining Data
  - *Census*
  - *Sampling*
  - *Experimentation*
  - *Observational Study*
- Census: A *census* occurs when ALL subjects/objects of the population are sampled.

  In other words, when taking a census, you are not collecting a sample; you are collecting information from EVERY element of the population.

- Any sample can be classified as one of the following:
  1. Nonprobability Samples
  2. Probability Samples
- *Nonprobability Samples* – the laws of probability are NOT used to obtain the samples.

  Listed below are some examples of nonprobability samples. This is by no means a complete listing of all different types of nonprobability samples, but it is a list of the ones more commonly used.

  - *Sample of Convenience* – the samples already exist.

    For example, since I teach a pre-med class, if my population is pre-med students, then I could use my class as a sample. This would be a sample of convenience.

  - *Haphazard Selection* – the researcher and the subjects casually meet.

    For example, if my population is shoppers at a suburban mall and I go to the mall and just casually approach people to survey them, then I have a haphazard selection.

  - *Judgment Sampling* – the researcher uses his/her own judgment in choosing the subjects.

    For example, suppose again my population is shoppers at a suburban mall and I go to the mall with the goal of surveying shoppers to determine their opinions on the new shopping hours. If I purposely only approach those whom appear to be "serious" shoppers and not just using the mall for exercise or entertainment, then I have used judgment sampling.

- *Expert Sampling* – an "expert" picks the subjects.

  For example, if, as the researcher, my population consists of dolphins and I let the biologist choose the sample, then I have an expert sample because the "expert" about dolphins chose the sample.

- *Quota Sampling* – subjects are chosen so as to satisfy certain quotas that match the population.

  A quota sample can be a probability sample if the subjects are randomly chosen, but I included it here because this is generally how you see a quota sample done.

  For example, suppose my population of Baylor students contains 22% freshmen, 20% sophomores, 19% juniors, 18% seniors, and 21% graduate students. If I choose my sample so that 22% of the sample are freshmen, 20% are sophomores, 19% are juniors, 18% are seniors, and 21% are graduate students, then I have a quota sample.

- Probability Samples – the sample is obtained by a chance process. In other words, the laws of probability are used to choose the sample.

  - *Simple Random Samples* – all samples of the same size have equal probability of being selected.

    One good analogy of obtaining a simple random sample is as follows: Suppose I want to randomly sample ten of my nursing students. To do this, I can put all their names in a sack, shake the sack, pull out a name, shake the sack again, pull out another name and continue this process until ten names have been drawn. Now in reality, we use a computer to randomly generate a sample.

    Question: Do you think it is possible to get a sample of 100 randomly chosen Baylor students that does not contain any seniors?

    Answer: Yes, it is possible. Not all random samples are well-representative samples of the population. When appropriate we can improve by performing one of the sampling schemes below.

  - *Stratified Random Samples* – population is divided into subpopulations (called strata) and a random sample is obtained from each strata.

    For example, suppose I want to obtain a sample of health education majors. To insure that I get all student classifications in my sample, I can divide my population of health education majors into five stratas: freshman, sophomore, junior, senior, and graduate student. Then I would take a random sample from each stratum to obtain a stratified random sample.

  - *Cluster Samples* – population is once again divided into stratas or clusters. The researcher randomly selects some of the strata or clusters and selects ALL subjects from the chosen strata. (Also when the subjects are naturally clustered together).

    For example, suppose I want to sample residents of a very large apartment complex that has units that are broken into "groups" (clusters) of six apartments. To obtain a cluster sample, I would randomly choose the "groups" (clusters) to sample and then sample every apartment unit in that chosen group.

    Another example: Suppose my population is tires. Since tires are clustered on vehicles, I would randomly choose the car and then sample each tire on the car.

  - *Systematic Sampling* – choose a random starting point from a list of subjects and then select every n^{th} subject on the list.

    For example, suppose I want to sample ten patients at General Mercy Hospital. Suppose the administrator at General Mercy Hospital has a numbered list of ALL 100 patients currently in the hospital. If I randomly choose a starting number between one and ten, say three, and then choose every tenth one, the third, thirteenth, twenty-third, etc., then I have a systematic sample.

Introduction to Statistics **7**

- Sampling Error: the extent to which the value of a statistic may differ from the parameter it predicts because of the way results vary from sample to sample. For example, suppose the true value of the parameter is 56%, but we predict 53%. Then the sampling error is -3%.

Read the following case study that illustrates some problems with sampling:

▶ Case Study 2: The 1936 Literary Digest Poll:

The U.S. presidential election of 1936 pitted Alfred Landon, the Republican governor of Kansas, against the incumbent President, Franklin D. Roosevelt. At the time of the election, the nation had not yet emerged from the Great Depression, and economic issues such as unemployment and government spending were the dominant themes of the campaign.

The *Literary Digest*, one of the most respected magazines of the time, conducted a poll a couple of weeks before the election. The magazine had been polling the electorate since 1916, always accurately predicting the results of the election. Based on its 1936 poll, the *Literary Digest* predicted that Landon would get 57% of the vote against Roosevelt's 43%. The actual results of the election were 62% for Roosevelt against 38% for Landon. The difference between the poll's prediction and the actual election results was a whopping 19%, the largest error ever in a major public opinion poll.

Ironically, the 1936 *Literary Digest* poll was one of the largest and most expensive ever conducted, based on a huge sample of approximately 2.4 million people. For the same election, a Gallup poll based on a much smaller sample of approximately 50,000 people was able accurately to predict a victory for Roosevelt. What happened?

The sample for the *Literary Digest* poll was chosen by putting together, in one enormous list, the names of every person listed in a telephone directory anywhere in the United States, as well as the names of people on magazine subscription lists and rosters of clubs and professional associations. Altogether, a mailing list of about 10 million names was created. Every name on this list was mailed a mock ballot and asked to mark it and return it to the magazine.

One cannot help but be impressed by the sheer ambition of such a project. And it is not surprising that the magazine's confidence in the results was in direct proportion to the magnitude of the effort. In its issue of August 22, 1936, the *Literary Digest* crowed:

Once again, [we are] asking more than ten million voters—one out of four, representing every county in the United States—to settle November's election in October.

Next week, the first answers from these ten million will begin the incoming tide of marked ballots, to be triple-checked, verified, five-times cross-classified and totaled. When the last figure has been torted and checked, if past experience is a criterion, the country will know to within a fraction of 1 percent the actual popular vote of forty million [voters].

When reality hit, it hit hard. Soon after the election, with its credibility badly damaged and its sales drying out, the *Literary Digest* magazine went out of business, the victim of a statistical *faux-pas*.

The first thing seriously wrong with the *Literary Digest* poll was in the selection process for the names on the mailing list. Names were taken from telephone directories, rosters of club members, lists of magazine subscribers, and the like. Such a list was inherently slanted toward members of the middle and upper

TANNEBAUM, PETER, EXCURSIONS IN MODERN MATHEMATICS, 5th edition, © 2004. Reprinted by permission of Pearson Education, Inc., Upper Saddle River, NJ.

8 Introduction to Statistics

classes. Telephones in 1936 were something of a luxury. So, too, were club memberships and magazine subscriptions, at a time when 9 million people were unemployed. At least with regard to *economic status*, the *Literary Digest* mailing list was far from being a representative cross section of population. This was a critical problem, because voters often vote on economic issues, and given the economic conditions of the time, this was especially true in 1936.

When the choice of the sample has a built-in tendency (whether intentional or not) to exclude a particular group or characteristic within the population, we say that a survey suffers from **selection bias**. It is obvious that selection bias must be avoided, but it is not always easy to detect it ahead of time. Even the most scrupulous attempts to eliminate selection bias can fall short (as will become apparent in our next case study).

The second problem with the *Literary Digest* poll was that out of the 10 million people whose names were on the original mailing list, only about 2.4 million responded to the survey. Thus, the size of the sample was about one-fourth of what was originally intended. When the proportion of respondents to the total number of people intended to be in the sample,[2] called the **response rate**, is low, a survey is said to suffer from **nonresponse bias**. For the *Literary Digest* poll the response rate was 24%, which is extremely low.

It is well known that people who respond to surveys are different from people who don't, not only in the obvious way (their attitude toward the usefulness of surveys) but also in more subtle ways: They tend to be better educated and in higher economic brackets (and are in fact more likely to vote Republican). Thus, nonresponse bias is a special type of selection bias—it excludes from the sample reluctant and disinterested people. But don't we want them represented?

Eliminating nonresponse bias from a survey is difficult. In a free country we cannot force people to participate, and paying them is hardly ever a solution, since it can introduce other forms of bias. Some ways of minimizing nonresponse bias are known, however. The *Literary Digest* survey was conducted by mail. This approach is the most likely to magnify nonresponse bias, because people often consider a mailed questionnaire just another form of junk mail. Of course, considering the size of the sample, the *Literary Digest* really had no other choice but to use mailed questionnaires. Here we see how a big sample size can be more of liability than an asset.

Nowadays, almost all legitimate public opinion polls are conducted either by telephone or by personal interviews. Telephone polling is subject to slightly more nonresponse bias than personal interviews, but it is considerably cheaper. In some special situations, however, telephone polls can be so biased as to be useless.[3]

The *Literary Digest* story has two morals: (1) *you'll do better with a well-chosen small sample than with a badly chosen big one*, and (2) *watch out for selection bias and nonresponse bias*.

- The *Literary Digest* illustrated certain problems with sampling, i.e., biases that can occur.

  Biases can cause a researcher to draw incorrect conclusions!

- *Selection Bias* (also called undercoverage) occurs when one excludes a specific characteristic or segment of the population.

  For example, in the *Literary Digest* Poll, members of the lower to middle class were excluded because they did not have phones, jobs, or subscribe to magazines or they were not members of professional organizations.

- *Response Bias* occurs when subjects respond incorrectly by lying or exaggerating or forgetting or not understanding the question. The question may also be a misleading question.

Introduction to Statistics **9**

For example, if you ask individuals if they have ever taken illegal drugs, they may not admit to doing so.

- *Nonresponse Bias* occurs when the response rate is low.

  The type of people who will respond to a survey is very different from the type of people who will not respond; that difference can significantly affect the results of the survey. The *Literary Digest* Poll suffered from Nonresponse Bias since the response rate was only 24%.

- There are two "types" of Experiments:

  1. Observational Study – An observational study occurs when the researcher simply observes the subjects in a specific situation or "natural habitat" or when simply comparing two or more groups. In an observational study, the subjects are not sampled nor included in an experiment. They are simply observed in their "natural habitat" or data is simply compared. An observational study can only show association, not causation. In other words, you cannot show causation with an observational study!

  2. Designed Experiment – has three basic principles:

     a. Control

     b. Randomization – randomly divides subjects into groups

     c. Replication – have sufficient number of subjects used to have groups similar in all aspects except the treatment

- *Treatment Group* – the group that receives the treatment.

- *Control Group* – group that does not receive the treatment.

- *Confounding factor* – anything that affects the outcome of the experiment other than the treatment. Also called an extraneous factor.

- *Blind Study* – when the subjects don't know if they are receiving the treatment or a placebo

- *Double-Blind Study* – when neither the subjects nor the doctors know if subjects are in the treatment or control group.

- The four basic experimental designs

  - Completely Randomized Design

  - Randomized Block Design

  - Latin Square Design

  - Factorial Design

- Problems with Experiments

  - *Placebo Effect* – when subjects improve because they believe they are receiving the treatment

  - *Rosenthall Effect* – when the researcher or experimenter unintentionally influences the outcome through facial expressions, body language, or voice.

  - *Hawthorne Effect* – when the outcome is affected because people change how they behave because they know they are being watched.

  - *Missing Data*

10   Introduction to Statistics

➡ Case Study 5. The 1954 Salk Polio Vaccine Field Trials:

Polio (infantile paralysis) has been practically eradicated in the western world. In the first half of the twentieth century, however, it was a major public health problem. Over one-half million cases of polio were reported between 1930 and 1950, and the actual number may have been considerably higher.

Because polio attacks mostly children and because its effects can be so serious (paralysis or death), eradication of the disease became a top public health priority in the United States. By the late 1940s it was known that polio is a virus and as such can best be treated by a vaccine which is itself made up of a virus. The vaccine virus can be a closely related virus that does not have the same harmful effects, or it can be the actual virus that produces the disease but which has been killed by a special treatment. The former is known as a *live-virus vaccine*, the latter as a *killed-virus vaccine*. In response to either vaccine the body is known to produce *antibodies* which remain in the system and give the individual immunity against an attack by the real virus.

Both the live-virus and the killed-virus approaches have their advantages and disadvantages. The live virus approach produces a stronger reaction and better immunity, but at the same time it is also more likely to cause a harmful reaction and in some cases even to produce the very disease it is supposed to prevent. The killed-virus approach is safer in terms of the likelihood of producing a harmful reaction, but it is also less effective in providing the desired level of immunity.

These facts are important because they help us understand the extraordinary amount of caution that went into the design of the study that tested the effectiveness of the polio vaccine. By 1953, several potential vaccines had been developed, one of the more promising of which was a killed-virus vaccine developed by Jonas Salk at the University of Pittsburgh. The killed-virus approach was chosen because there was a great potential risk in testing a live-virus vaccine in a large-scale study, and a large-scale study was needed to collect enough information on polio (which in the 1950s had a rate of incidence among children of about 1 in 2000).

The testing of any new vaccine or drug creates many ethical dilemmas which have to be taken into account in the design of the study. With a killed-virus vaccine the risk of harmful consequences produced by the vaccine itself is small, so one possible approach could have been to distribute the vaccine widely among the population (ideally giving it to every child, but this was not possible because supplies were limited) and then follow up on whether there was a decline in the national incidence of polio in subsequent years. This is called the *vital statistics* approach and is the simplest way to test a vaccine. This is essentially the way the smallpox vaccine was determined to be effective. The problem with such an approach for polio is that polio is an epidemic type of disease, which means that there is a great variation in the incidence of the disease from one year to the next. In 1951, there were close to 60,000 reported cases of polio in the United States, but in 1952 the number of reported cases had dropped to almost half (about 35,000). Since no vaccine or treatment was used, the cause of the drop was the natural variability that is typical of epidemic diseases. So, if a totally ineffective polio vaccine had been tested in 1951 without a control group, the observed effect of a large drop in the incidence of polio in 1952 could have been interpreted as a proof that the vaccine worked.

The final decision on how best to test the effectiveness of the Salk vaccine was left to an advisory committee of doctors, public officials, and statisticians convened by the National Foundation for Infantile Paralysis and the Public Health Service. In order to isolate the cause under investigation (the Salk vaccine)

TANNEBAUM, PETER, EXCURSIONS IN MODERN MATHEMATICS, 5th edition, © *2004. Reprinted by permission of Pearson Education, Inc., Upper Saddle River, NJ.*

from other possible causes of the desired effect (a reduction in the incidence of polio), it was decided that the study would be controlled, with a treatment group (those receiving the actual vaccine) and a control group (those receiving a **placebo**—in this case a shot of harmless salt solution).

The reasons for using placebos in controlled studies go back to our desire that the treatment and control groups be as equal as possible in all respects, except of course that one group is receiving the vaccine and the other one isn't. It is a well-known fact that the mere *thinking* that one is getting a helpful vaccine or pill can actually produce positive results. Placebos are the standard way of eliminating this confounding variable known as the *placebo effect*.

It goes without saying that in a controlled placebo study it is essential that neither the members of the treatment group nor the members of the control group know to which of the two groups they belong. When this is the case the study is said to be a **blind study**. It is also desirable that the scientists conducting the study not know which subjects are taking the actual treatment and which are taking the placebo. The purpose is to make the observation, analysis, and interpretation of the results of the study as impartial as possible.

Making the Salk vaccine study double-blind was particularly important because polio is not an easy disease to diagnose—it comes in many different forms and degrees. Sometimes it can be a borderline call, and if the doctor collecting the data had prior knowledge of whether the subject had received the real vaccine or the placebo, the diagnosis could have been subjectively tipped one way or the other.

With all this background we can now describe the actual details of the experiment. Approximately 750,000 children were randomly selected to participate in the study. Of these, about 340,000 declined to participate and another 8500 dropped out in the middle of the experiment. The remaining children were divided into two groups—a treatment group and a control group—with approximately 200,000 children in each group. The choice of which children were selected for the treatment group and which for the control group was made by *random selection*.

Based on the data collected by the 1954 field trials, a massive inoculation campaign was put into effect. Today, all children are routinely inoculated against polio,[5] and polio has essentially been eradicated in the United States. Statistics played a key role in this important public health breakthrough.

Checking your Understanding: (Answers are found in the Appendix)

For the following scenario, determine the type of method or sampling scheme used to obtain the data.

a. John went to the food court of the mall and noted whether the patrons ate their purchased food at the food court or took their food to go.

b. Director Jones surveyed ten teachers from each school in Region 12 to determine how they feel about the new state requirements on teacher certification.

c. Director Smith randomly selected 20 schools from his district and surveyed every teacher in the selected school to see if they support the new requirements on student grading policies.

---

[5] The Salk vaccine, which had been used for many years, was replaced some years ago by the Sabin vaccine, an oral vaccine based on the live-virus approach.

Section 2: Homework:

1. For each scenario below, choose the sampling scheme or method that was used to obtain the data.

   a. Jane randomly sampled 200 High School students at Richman High to get their opinion on the new dress code.

   b. Sarah's company is divided into departments. She randomly selected 10 departments and interviewed every person working in that department asking their opinion on the new company parking policy.

   c. George sampled every 10th person on the registrar's list about the new required Freshman Seminar.

   d. Dean Jones randomly surveyed 100 students from every grade classification to get their opinion of the new COVID screening required for all the students to attend class on campus.

   e. In order to determine how the new city ordinances with respect to playground equipment affected how children play, a child psychologist visited the local playground and watched the children playing on the playground.

   f. To determine how the members of his church feel about the new Sunday School hours, the pastor mailed a survey to every member.

2. For each scenario below, determine the bias or the effect that is being described.

   a. Students were asked if they ever purchased a term paper to turn in for a grade that was not written by them.

   b. Sarah went to the football game to survey students to see what all students of the university felt about the new parking restrictions on campus.

   c. James sent 1000 surveys to residents of his home town. He only received 3% of them back.

   d. Coach Willis noticed that his students did the exercises better when either he or his assistant was present in the weight room.

   e. Researchers discovered that patients reported improvement in migraine pain even though nothing more was done during the procedure except a simple saline shot.

   f. The children that participated in the study changed their answers to the questions whenever the researcher looked surprised by the child's original response.

➡ Section 3: Descriptive Statistics – Organizing Data:

Once the data has been collected, it is important to organize the data in a table or graph. The goal of organizing data is to make the data more "comprehendible." If you simply list out the data in rows of numbers, then there is little meaning to the data; it is simply a row of numbers. So we group the data into tables or make graphs (a picture) of the data. The type of graph created varies depending upon if the data is quantitative or qualitative. For quantitative data, the type of table or graph is also dependent upon if the data is discrete or continuous. Some common tables and graphs are listed below.

Tables	Graphs
Frequency Distribution	Dot Plot
Grouped Frequency Distribution	Histogram
Relative Frequency Distribution	Bar Graph
	Pie Chart

Introduction to Statistics 13

- The following tables are typically used for quantitative data:
  - Frequency Distribution: a table that lists the data values and the number of times the data value occurs. (The number of times a data value occurs is called its frequency.) This table can be used for discrete or continuous data.

    Example: Suppose the following frequency distribution table represents the age at which a child sampled first loses a tooth.

X	3	4	5	6	7
$f(x)$	1	5	10	6	3

  - The symbol, X, is used to represent the unique data values.
  - The notation, $f(x)$, is used to represent the frequency of the data.
  - How many children sampled were five years old when they first lost a tooth?

    Answer:

    From the table we see that $f(x) = 10$ when $x = 5$. Thus, 10 children were five years old when they first lost a tooth.

  - What is the value n?

    Answer:

    Recall n is the sample size. Thus, $n = 1 + 5 + 10 + 6 + 3 = 25$.

- Dot Plot: A "picture" of a frequency distribution commonly used is the Dot Plot. In a Dot Plot, dots are used to represent the frequency of the data.

  Example:

14   Introduction to Statistics

- Relative Frequency Distribution: a table that lists the data values and the percentage of times the data value occurs. (The percentage of times a data value occurs is called its *relative frequency*).

Example: Consider again the age at which a sampled child first loses a tooth. The frequency is now listed in terms as a percentage instead of as a count.

X	3	4	5	6	7
f(x)	0.04	0.20	0.40	?	?

- Again, the symbol, X, is used to represent the unique data values.
- The notation, $f(x)$, is now used to represent the relative frequency of the data.

What is the relative frequency for the data value, $X = 6$?

Answer:

Using the frequency table on the previous page, relative frequency = percentage = 6/25 = 0.24

What is the relative frequency for the data value, $X = 7$?

Answer:

Relative frequency = percentage = 3/25 = 0.12

What is the sum of the relative frequencies?

Answer: 1

- Grouped Frequency Distribution. When the sample size is large, the data is typically combined into groups. Recall that the goal of organizing the data into tables and graphs is to aid in understanding the meaning of the data. If the frequency distribution becomes too large, the "meaning" of the data is difficult to "see," thus the data is combined into groups.

The groups can be determined by "natural" grouping in the data or by the sample size, n. For example, if the data obtained is grades, these are naturally grouped in terms of As, Bs, etc. If the data does not naturally fall into groups, the size of the data set, n, determines the number of groups used. A good rule of thumb for determining the number of groups in a grouped frequency distribution is $\sqrt{n}$. A requirement of a grouped frequency distribution is that every data value should belong to a group and the first and last group should contain a data value.

(I am not going to spend time teaching you how to create a grouped frequency distribution by hand. The main thing is to understand what a grouped frequency distribution is and how to read one.)

- Histogram – a "picture" of a grouped frequency distribution is called a histogram. The horizontal axis represents the grouped data. The height of the bar drawn represents the frequency or relative frequency of the data. The width of the bar equals the width of the group. For example, if grouping by grade, the points 90 – 100 represent an A, so the width of the bar equals ten. For data that is not grouped, the width of the bar is determined by the number of data values and the number of groups chosen.

(Again, I am not going to spend time in class teaching you how to create a histogram by hand. We will use technology to create histograms.)

Introduction to Statistics   **15**

Example: Suppose again we are interested in the age at which a child first loses a tooth. To be measured more exactly, the data set takes into account a more exact age. Instead of just putting 5 for five years, it includes the months as well. For example, the value 5.25 implies the child was five years and three months old.

> We will use JMP to create the histogram for us. Listed below are the steps:
>
> 1. Under *File*, choose *New*, then choose *Data Table*. Click on the first cell of column 1 to enter the unique data values. Choose *New Column* to get column 2. Click on the first cell of column 2 to enter the frequency of the data.
> 2. Under *Analyze*, choose *Distribution*. Click on the column containing the unique data and click on the *Y, Columns* button. Click on the column containing the frequencies and click on the *Freq* button. Click *Ok*.
> 3. Under the red triangle, choose *Display Options*, then choose *Stack*.
> 4. Under the red triangle, choose *Histogram Options*, then choose *show counts* to show the counts or choose show percents to get the percentages.
> 5. To add the label, place cursor on the *x*-axis and right click, choose add axis level.
> 6. Once a label is added, you can edit it by clicking on it.

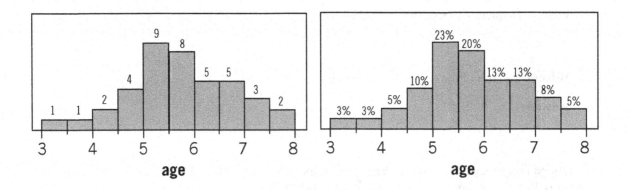

- Another graph used for quantitative data, discrete or continuous, is called a stem-n-leaf plot:

To get the stem-n-leaf plot in JMP, simply choose it as a choice under the options.

Stem	Leaf	Count
7	56	2
7	122	3
6	56689	5
6	22333	5
5	56788889	8
5	122233344	9
4	8899	4
4	13	2
3	8	1
3	1	1

3|1 represents 3.1

16   Introduction to Statistics

- The following graphs are typically used for qualitative data:
  - Bar graph: In a bar graph, a bar is used to represent each "category of the data." The height of the bar represents the total percentage or the total count of how many fall in the specific category. With a bar graph, the bars are not touching.

> To create the bar graph using JMP use the following steps:
> 1. Once the data is inputed, under the graph menu, choose *Graph Builder*.
> 2. Click on the column containing the unique data and drag to the *x*-axis. At the top of the chart, click on the "Histogram" button.
> 3. To obtain the percentages, click on *Summary Statistics*, and choose *% of total*.

  - Pie chart: In a pie chart, each category is represented by a slice of the pie. The size of the pie slice represents the total percentage for each category. For a pie chart, we can change the color of the slices to represent the categories.

> To create the pie chart using JMP use the following steps:
> 1. Once the data is inputed, under the *Graph*, choose *Graph Builder*.
> 2. Click on the column containing the category and drag to the *x*-axis. At the top of the chart, click on the "Pie Chart" button.
> 3. On the left side, there is an options choice. Change this from bar graph to pie chart.

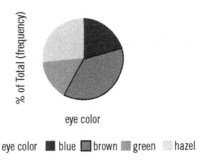

Introduction to Statistics

➡ Section 4: Descriptive Statistics – Describing Data

When describing the data, the goal is to calculate a single value to describe a certain aspect of the data. There are three different aspects of the data we like to describe: the center, the variability, and the location.

Measures of Central Tendency – conveys a "typical" value of the data set. There are three measures of central tendency:

1. Mean – the center of the data

2. Median – the middle of the data

3. Mode – the data value that occurs the most often

- The mean is the true center of the data. It is the balancing point of the data. Graphically, it is the point at which the graph would be balanced. The deviation above the mean will always equal with the deviation below the mean.

Let X = random variable.

There are two different means:

   1. The population mean, denoted as _____.

   2. The sample mean, denoted as _____.

The Population Mean:

$$\mu = \frac{\sum\limits_{i=1}^{N} X_i}{N}$$

- to find μ add up all the data values in the population and divide by N.

Is μ a parameter or a statistic?

Is μ a constant or a random variable?

μ is _____ unless a _____ is performed.

Can we calculate probabilities with respect to μ?

Answer:

18 Introduction to Statistics

The Sample Mean: $\bar{x} = \dfrac{\sum_{i=1}^{n} X_i}{n}$

- to find $\bar{x}$ add up all the data values in the sample and divide by n.

Is $\bar{x}$ a parameter or a statistic?

Is $\bar{x}$ a constant or a random variable?

$\bar{x}$ is _____ . It is calculated from the _____ .

Can we calculate probabilities with respect to $\bar{x}$?

Answer:

** We use $\bar{x}$ as an estimate for μ.

- Properties of the mean:
  - Most commonly used.
  - May not be a value in the data set.
  - Is not a resistance measure, which is a disadvantage to the mean. (We will talk about this more later.)
  - We do not round the mean to a whole number, even if the element in the population can't be a whole number. For example, the average number of kids per family is 1.7. Even though you cannot have 0.7 of a child, we don't round the mean to 2; we keep it as 1.7.

The Median:

- The median is the value such that 50% of the data values are less than that value and 50% of the data values are greater than that value. The median divides the data into half.
- *Notation*: M = the population median

  $\tilde{x}$ = the sample median

To find:

1. Rank the data (i.e., put the data in order)
2. Find the location of the data using the formula $(n+1)/2$.

Introduction to Statistics **19**

If there are an even number of data values, then the location value above will not be an integer. So to find the median, you average the two middle values.

For example, if the data set is

$$2 \qquad 6 \qquad 8 \qquad 10$$

Then the location of the median is $(4 + 1)/2 = 2.5$, and thus the median is 7

If there are an odd number of data values, then the location value above will be an integer. So to find the median you just determine the data value in that location.

For example: if the data set is

$$2 \qquad 8 \qquad 10 \qquad 14 \qquad 18$$

Then the location of the median is $(5 + 1)/2 = 3$, and thus the median is 10.

- Properties of the median:
  - May not be a value in the data set.
  - Is a resistance measure—which is an advantage of the median.

Definitions

Outlier – a data value that is an extreme value

Resistance Measure – a measure that is not affected by outliers

- Example: Consider the following data set

$$46 \qquad 44 \qquad 43 \qquad 38 \qquad 45$$
$$42 \qquad 45 \qquad 44 \qquad 46 \qquad 44 \qquad 135$$

- Note that the sample mean is $\bar{x} = 52$ and the sample median is $\tilde{x} = 44$.
  Which measure do you think best represents the typical value in the data set?
  Answer:

  Why is there such a difference between the mean and the median?
  Answer:

  Does changing 135 to 48 affect the mean?
  Answer:

Does changing 135 to 48 affect the median?
Answer:

Suppose the data set above represents the average starting salaries (in thousands) for nurses.

If, as the researcher, I report that the average starting salary is 52,000, am I correct?

Answer:

Am I being misleading?

Answer:

It would be better for me to report that the _____ starting salary is _____.

Moral of the story: a resistance measure is a measure that is not affected by outliers. The

_____ is a resistance measure. The _____ is not a resistance measure.

When the data set contains outliers, the _____ should be used as the measure of

central tendency.

More on Outliers:

Recall that outliers are data points that are much larger or much smaller than the other data values, i.e., an extreme value. However, the "extremes" are not necessarily always outliers. There are different methods for determining if an extreme value is an actual outlier. We will discuss these later.

Outliers can sometimes lead to complications in data analysis and erroneous conclusions if not treated appropriately. For example, when calculating the average length of stay in the county jail for individuals charged with a class C felony during a particular year, an analyst discovers that all individuals arrested spent just a few days in jail before appearing before a judge except for one individual who stayed for 25 months. This one value made the average length of stay to be six weeks.

Is the value six weeks truly typical of how long individuals were staying in jail before seeing a judge?

Answer:

Introduction to Statistics **21**

Possible reasons for outliers:

1. The researcher sampled from a "different" population without realizing it. The outlier should not be discarded, but if it can be shown that they sampled from the wrong population, then the values will not be used in the original study.

   For example, in Australia researchers were investigating opossums and realized that every once in awhile, there would be an outlier for the tail, ring finger, or ear measurement. Fortunately they did not throw away these outliers and soon saw a pattern in the outliers. From the outliers, they discovered a whole new species of opossum; one that had a longer tail, shorter finger, and shorter ears than the opossums they were studying.(From Chance magazine)

2. A mistake was made while taking a measurement or entering it into the computer. After certain justification, the value should be either corrected or removed from the data set.

   For example, a common measurement error for newborn babies is their length. This is due to the fact that different tape measures have different "starting points" and the nurse may not realize this. Also, newborn babies are sometimes difficult to "stretch out" to measure. (When the author's oldest daughter Ana was born, her length was measured as 22 inches. However for her first week check-up, she was measured as being 21 inches long.)

3. The researcher sampled from the "correct" population and no error was made. The outlier is a legitimate data value from the population and represents natural variability in the data. In this case, the data value should not be discarded because it provides important information about the population.

   For example, when the author's daughter Ana was born she weighed 11 lbs, 1 oz. She was not diabetic and neither was Ana. Ana was simply just a big baby!

Outliers should never be discarded without proper justification!

<u>The Mode</u>:

- May not be unique
- The mode is less affected by outliers than either the mean or median. Unfortunately, the mode will vary more from sample to sample (from the same population) than either the mean or the median, making it a poor choice as a measure of central tendency.
- <u>Measures of Dispersion</u>

1. Do the following data sets have the same mean?
2. Are they similar data sets?

As illustrated above, two data sets can have the same "typical value", but be completely different data sets. Thus, in addition to calculating a Measure of Central Tendency, a Measure of Dispersion should be calculated as well.

- A measure of dispersion conveys information about the amount of variability in the data. There are four different measures of dispersion we will discuss.

  1. Range
  2. Variance
  3. Standard Deviation
  4. Interquartile Range

The Range:

- The range measures the variability by comparing the largest and smallest data value.

  $R = x_L - x_S$ (i.e., the largest data value minus the smallest data value)

  Is the range a resistance measure?

  Answer:

The Variance:

- The variance measures the variability by comparing each data value to the mean. Just like with the mean there are two variances:

  1. The population variance, denoted as _____.
  2. The sample variance, denoted as _____.

The Population Variance: $\sigma^2 = \dfrac{\sum_{i=1}^{N} (x_i - \mu)^2}{N}$

Is σ^2 a parameter or a statistic?

Is σ^2 a constant or a random variable?

The Sample Variance: $s^2 = \dfrac{\sum_{i=1}^{n}(x_i - \bar{x})^2}{n-1} = \dfrac{n\left(\sum x_i^2\right) - \left(\sum x_i\right)^2}{n(n-1)}$

Is s^2 a parameter or a statistic?

Is s^2 a constant or a random variable?

Introduction to Statistics **23**

Interpreting the variance: The larger the number, the more variability there is in the data set.

- Is the variance a resistance measure?

  Answer:

Example of finding s^2: Suppose the data set is as follows: 1   2   3.

then $s^2 = \dfrac{3(1^2 + 2^2 + 3^2) - (1 + 2 + 3)^2}{3(2)} = \dfrac{3(14) - 6^2}{6} = \dfrac{42 - 36}{6} = 1$

The Standard Deviation:

- The standard deviation measures the variability in the original units of the data

  $\sigma = \sqrt{\sigma^2}$ σ is the population standard deviation

  $s = \sqrt{s^2}$ s is the sample standard deviation

  - Is the standard deviation a resistance measure?

    Answer:

- Coefficient of Variation. When wanting to compare the dispersion (or variability) between two sets of data one needs to be careful because:

  1. The data might have different units. For example, one might be comparing hemoglobin levels measured in gm/dL versus body weight measured in pounds.

  2. The data might have extremely different means. For example, one might be comparing weights of babies versus weights of adults.

- To compare the variability of two data sets, you can calculate the Coefficient of Variation:

  $C.V. = \dfrac{s}{\bar{x}}(100)$

  The *C.V.* is given as a percentage. The data set with the largest *C.V.* has the most variability.

- Measures of Location - conveys information about the location of a specific data value compared to the other data values.

  1. Percentiles

  2. Quartiles

  3. Standardized Values

- Percentiles: The mth percentile, P_m, is the value x such that $m\%$ of the data values are less than x and $(100 - m)\%$ are greater than x.

If you have your IQ tested, would you rather be told you are in the 98th percentile or the 2nd percentile?

Answer:

• Quartiles: Quartiles are specific percentiles. The Quartiles divide the data into quarters.

$Q_1 = P_{25}$

$Q_2 = P_{50} = $ _____

$Q_3 = P_{75}$

There are many different methods for finding percentiles/quartiles, which will result in different answers. For this class, we will not compute the percentiles/quartiles by hand. We will use JMP to calculate. We will also use JMP to calculate the mean and variance.

To obtain the mean, standard deviation, quartiles and the following percentiles: $P_{0.5}$, $P_{2.5}$, P_{10}, $P_{97.5}$, $P_{99.5}$, perform the following steps:

1. Enter the data. Under the *Analyze* menu, choose *Distribution*.

2. Click on the column containing the data, click on *Y*, hit *OK*.

3. (To obtain the other measures discussed above, click on the red triangle next to *Summary Statistics*, choose *Customize Summary Statistics*.)

To obtain a specific percentile, do the following:

1. Under the *Col* menu, choose *New Column*.

2. Under *Column Properties*, choose *Formula*.

3. Choose *Statistical* and then choose *col(Quantile)*.

4. Click on column containing data and click on first box.

5. In second box, type in the desired percentile as a decimal.

6. Hit *Ok*.

Let's illustrate the above steps with the following example:

• Suppose the following data represents the age at which a randomly selected American adult first received a speeding ticket.

18	18	19	19	20	21	21	21
22	22	23	23	23	32	32	32
32	33	36	36	37	39	39	40
41	42	42	43	44	44	45	46
48	49	50	51	52	56	56	61
63	65	88					

Introduction to Statistics 25

Use JMP to find the requested percentiles and the quartiles (and moments):

JMP output:

Distributions age	Moments
Mean	38.232558
Std Dev	15.502438
Std Err Mean	2.3640996
Upper 95% Mean	43.003504
Lower 95% Mean	33.461612
N	43
Sum Wgt	43
Sum	1644
Variance	240.32558
Skewness	0.7913035
Kurtosis	1.0041674
CV	40.547739
N Missing	0

Distributions	age	Quantiles
100.0%	maximum	88
99.5%		88
97.5%		85.7
90.0%		59
75.0%	quartile	48
50.0%	median	39
25.0%	quartile	23
10.0%		19.4
2.5%		18
0.5%		18
0.0%	minimum	18

- Find and interpret the 10th percentile, P_{10}:

  Answer:

- Find the Quartiles:

  Answer:

Recall, when discussing Measures of Dispersion, we did not define the Interquartile Range. Now, we can.

- *The Interquartile Range*:

$$IQR = Q_3 - Q_1$$

The *IQR* describes the variability of the middle 50% of the data values.

The IQR for our speeding ticket example: IQR = 48 – 23 = 25

Interpretation:

Is the *IQR* a resistance measure?

Answer:

What Measure of Dispersion should we use if our data value contains outliers?

Answer:

What Measure of Dispersion should we use if our data set does not contain outliers? Why?

Answer:

Checking your Understanding: (Answers are found in the Appendix)

Use the following scenario to answer the questions below:

Suppose the average age of a women diagnosed with breast cancer is 62 years old with standard deviation 2.4 years. Suppose Dr. Veen, a well-known oncologist, has treated 100 women in his career. His records show that the average age of his patients is 59 years old with standard deviation 1.8 years.

a. What is the value of $\bar{x}$?

b. What is the value of σ?

c. Which measure describes the typical value of the population?

d. Which measure describes how much variability there is about the sample mean?

Introduction to Statistics **27**

Section 4: Homework:

1. For the following scenarios, choose which measure should be calculated. Be Specific.
    a. The goal is to describe the "typical" value in the data set and the data set contains an outlier.
    b. The goal is to describe the amount of variability in the data set and the data set is symmetric.
    c. The goal is to describe the true center, the balancing point of the data and the data set is symmetric.
    d. The goal is to describe the amount of variability in the data set and the data set is skewed right.

2. Suppose the following data set gives the age a person was the first time they received a winning lottery ticket. Use JMP to answer the questions below. Make sure and print your output!

    | 35 | 43 | 16 | 30 | 20 | 32 | 25 | 26 | 32 | 30 | 18 | 35 |
    | 21 | 27 | 34 | 5 | 38 | 31 | 38 | 41 | 15 | 22 | 28 | 38 |

    a. Find P_{15}.
    b. In the context of this problem, interpret P_{15}.
    c. Find the Quartiles.
    d. Find the Interquartile Range.
    e. Interpret the Interquartile Range.

3. Suppose you have the following data set

    | 23 | 27 | 31 | 29 | 32 | 99 |

    a. Which measure should you calculate to describe the typical value? Why?
    b. Which measure should you calculate to describe the amount of variability? Why?

4. True or False: $P(\mu > 1.2) = 0.1245$ is a valid probability statement.

➡ Section 5: Distribution Shapes:

When analyzing data, it is important to determine the shape of the data, since the shape can affect which measure we choose to describe the data and which method we use to analyze the data.

We determine the shape of the distribution by visualizing the histogram smoothed out into a curve, as shown in the graph below.

- distribution shapes
    - frequency "polygons" or "ogives"

28 Introduction to Statistics

- unimodality and skewness

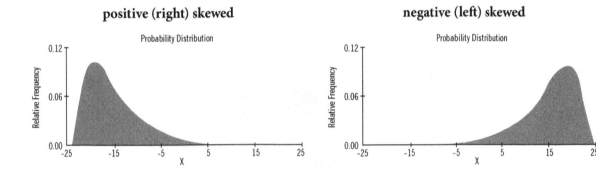

- As we will see later, we like the distribution of our data to be symmetric and not skewed.

- If the data is skewed, we should use the _____ as the measure of central tendency. Why?

  Answer:

- If the data is skewed, we should use the _____ as the measure of dispersion. Why?

  Answer:

- If the data is symmetric, we should use the _____ as the measure of central tendency. Why?

  Answer:

- If the data is symmetric, we should use the _____ as the measure of dispersion. Why?

  Answer:

More shapes:

- multimodal

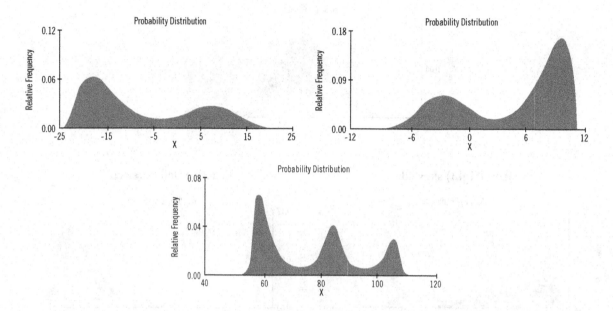

- Typical relationships among mean, median, and mode
    a. Positive (right) skewed   <u>usually</u> ↦ mode < median < mean
    b. Negative (left) skewed   <u>usually</u> ↦ mean < median < mode
    c. Unimodal
- Exceptions to the mean, median, mode inequalities
    - bimodal distributions
    - This distribution is approximately symmetric but not unimodal

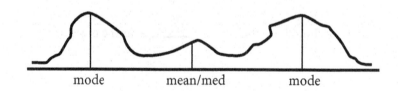

- It is easy to construct a distribution which violates a or b:

30   Introduction to Statistics

Section 5: Homework:

1. For the following histograms, determine the following:
   i. The shape of the distribution.
   ii. Whether $\bar{x} < \tilde{x}$ or $\bar{x} > \tilde{x}$ or $\bar{x} \approx \tilde{x}$.
   iii. Which Measure of Central Tendency you should calculate.
   iv. Which Measure of Dispersion you should calculate.

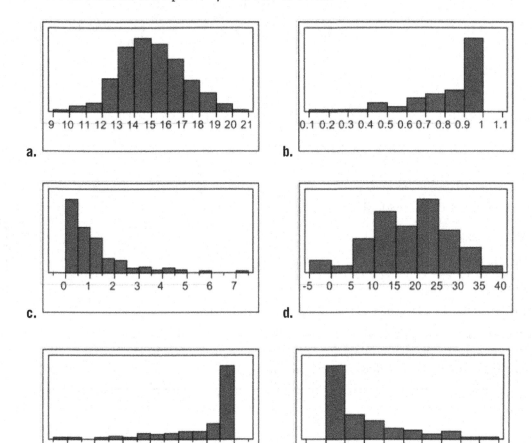

➡ Section 6: Modified Boxplots:

A "picture" of the data that displays the shape, the middle, the variability of the data and whether the data contains outliers. To find, we need to calculate the following:

1. The Quartiles – Q_1, Q_2, and Q_3
2. IQR
3. $Q_1 - 1.5*IQR$
4. $Q_3 + 1.5*IQR$

} These values are used for detecting outliers. Any value less than $Q_1 - 1.5*IQR$ is a possible outlier. Any value greater than $Q_3 + 1.5*IQR$ is a possible outlier.

Introduction to Statistics 31

- A boxplot:

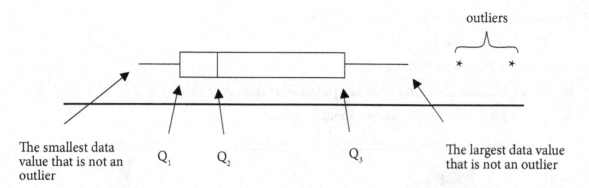

What does the "box" represent?

Steps for using JMP to find the Modified Boxplot:

1. Enter the data in the column.
2. Under the *Graph* menu, choose *Graph Builder*.
3. Click on the column containing the data and drag to the *x*-axis.
4. At the top of the window, click on the *Modified Boxplot Button*.
5. To copy the boxplot, click on the done button. Then you can select and copy. (To create another boxplot on the same scale as the first, click on the second data set and drag to the *x*-axis. Then click on the *Modified Boxplot Button* again.)

32   Introduction to Statistics

From the modified boxplot, we can describe the "typical" data value and the variability of the data using resistant measures.

We can discuss the quartiles and talk about the shape of the distribution. We can also determine at a glance if there are any outliers.

- Here are several examples contrasting box plots with underlying shapes

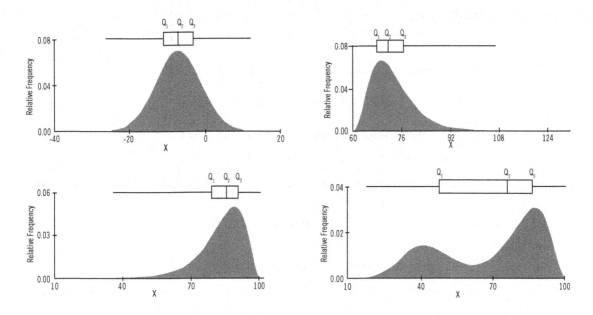

Section 6: Homework

1. Suppose the following data set gives the age a person was the first time they received a winning lottery ticket. Use JMP to answer the questions below. Make sure and print your output!

    | 35 | 43 | 16 | 30 | 20 | 32 | 25 | 26 | 32 | 30 | 18 | 35 |
    | 21 | 27 | 34 | 5 | 38 | 31 | 38 | 41 | 15 | 22 | 28 | 38 |

    a. Construct a Boxplot (Modified Boxplot).

    b. Determine if the data set contains any outliers. If there are outliers, make sure and state which data value(s) are outliers.

    c. Find the typical data value of the data set. Explain why you chose that measure.

2. On the last day of class, two graduate statistics classes at the University of Texas at Arlington were given a survey. One class was made up of statistic majors, while the other was made up of non-statistic majors. One of the questions asked was, "During the semester, how many hours did you study per week, to the nearest hour?" Use the modified boxplots below to answer the questions that follow.

Introduction to Statistics 33

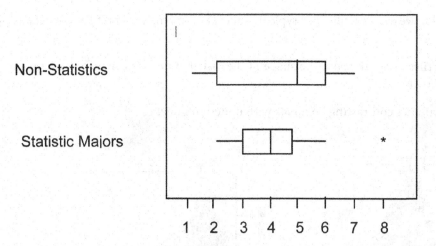

a. Estimate the median hours studied for non-statistic major students.

b. Which group has more variability in the middle 50% of the data?

c. Approximately what percent of non-statistic majors studied less than all the statistic majors sampled?

d. Is there an outlier present in either the statistic majors or the non-statistic majors? If so, what is the value of the outlier?

e. What is the approximate shape of the distribution for the non-statistic majors?

f. What is the number of hours such that only 25% of the non-statistic majors studied more than that per week?

g. What percent of the statistic majors studied less than four hours?

Section 7: Two Very Important Facts about Describing Data:

- Fact 1: When describing a data set you should always do the following three steps:

    1. Obtain a "picture" of the data
    2. Calculate a measure of central tendency
    3. Calculate a measure of dispersion.

    You should never describe a data set without doing all three steps or your description could be incorrect or at the least, very misleading.

- Fact 2: Be very careful looking at trends in averages of some quantity (like income) over time. What you may be observing is an increase in variability and skewness over time, resulting in an increase in averages, but leaving the modal values essentially unchanged. You would be able to detect that phenomenon if you had the three necessary tools mentioned above.

    For example suppose you have the graphs below that represent the waiting times at a doctor's office for three consecutive months. The graphs on the left describe the waiting times for Dr. Allison. The solid vertical line represents the averages. The graphs on the right describe the waiting times for Dr. Barnes. The solid vertical line represents the average and the dashed line represents the mode.

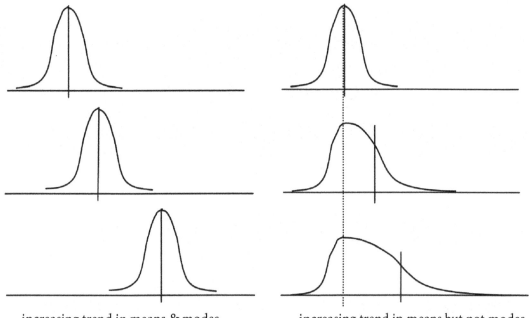

increasing trend in means & modes   increasing trend in means but not modes

If we just look at the solid vertical lines, how do the average waiting times compare for Dr. Allison and Dr. Barnes?

Answer:

What do the graphs on the left "tell us" about the overall waiting times for Dr. Allison?

Answer:

What do the graphs on the right "tell us" about the overall waiting times for Dr. Barnes?

Answer:

Taking into account the variability and the shape, can we conclude the waiting times for the two doctors are the same?

Answer:

Introduction to Statistics   35

➡ Chapter 1 Review Homework: _ _ _ _ _ _ _ _ _ _

1. Dr. Green is concerned about the sleeping habits of STA 1380 students. Altogether, there are five sections of STA 1380, each containing thirty students. Dr. Green decides to randomly select ten students from each of the five different classes. She finds that 38% of the sampled students sleep less than six hours per night on average. Upon hearing this, the head of the department is greatly concerned and checks the sleeping hours of all the STA 1380 students and finds only 13% are getting less than six hours of sleep per night.

   _____ **i.** The population in this study is _____.

        A. The fifty students surveyed.

        B. The ten students selected from Dr. Green's class.

        C. The entire student body of Baylor.

        D. All students currently enrolled in STA 1380.

   _____ **ii.** The sample in this study is _____.

        A. The fifty students surveyed.

        B. The ten students selected from Dr. Green's class.

        C. The entire student body of Baylor.

        D. All students currently enrolled in STA 1380.

   _____ **iii.** In the paragraph above, Dr. Green's initial finding of 38% is a(n) _____.

        A. Error.

        B. Statistic which is a random variable.

        C. Statistic which is a constant.

        D. Parameter which is a random variable.

        E. Parameter which is a constant.

        F. Random Variable.

   _____ **iv.** The head of the department's finding of 13% is a(n)_____.

        A. Error.

        B. Statistic which is a random variable.

        C. Statistic which is a constant.

        D. Parameter which is a random variable.

        E. Parameter which is a constant.

        F. Random Variable.

Introduction to Statistics 37

_____ **v.** What type of sampling scheme did Dr. Green use?

    A. Simple Random Sampling.

    B. Systematic Sampling.

    C. Stratified Sampling.

    D. Cluster Sampling.

    E. Sample of Convenience.

2. Mrs. Strong is concerned about the eating habits of the students at Crawford Elementary. Altogether, there are twenty different classes of kids, each containing thirty students. Mrs. Strong decides to randomly select twelve classes and sample each child in the sampled class. She finds that 48% of the sampled students ate a healthy breakfast before coming to school. Upon hearing this, the superintendent, Dr. Jones, is greatly concerned and checks the eating habits of all the Crawford Elementary students and finds that 63% ate a healthy breakfast before coming to school.

_____ **i.** The population in this study is _____.

    A. The 360 students surveyed.

    B. All the elementary students at Crawford Elementary.

    C. The 30 students selected from Mrs. Strong's class.

    D. The entire student body of Crawford ISD.

    E. None of the above.

_____ **ii.** The sample in this study is _____.

    A. The 360 students surveyed.

    B. All the elementary students at Crawford Elementary.

    C. The 30 students selected from Mrs. Strong's class.

    D. The entire student body of Crawford ISD.

    E. None of the above.

_____ **iii.** In the paragraph above, Mrs. Strong's initial finding of 48% is a(n) _____.

    A. Error.

    B. Statistic which is a constant.

    C. Statistic which is a random variable.

    D. Parameter which is a constant.

    E. Parameter which is a random variable.

    F. None of the above.

_____ **iv.** Dr. Jones's finding of 63% is a(n) _____.

    A. Error.

    B. Statistic which is a constant.

    C. Statistic which is a random variable.

    D. Parameter which is a constant.

    E. Parameter which is a random variable.

    F. None of the above.

_____ **v.** What type of sampling scheme did Mrs. Strong use?

    A. Simple Random Sampling.

    B. Systematic Sampling.

    C. Stratified Sampling.

    D. Cluster Sampling.

    E. Sample of Convenience.

3. Fill in the blank with the correct symbol.

  a. We denote the population mean with the symbol _____.

  b. We denote the sample mean with the symbol _____.

  c. We denote the population variance with the symbol _____.

  d. We denote the sample variance with the symbol _____.

  e. We denote the population size with the symbol _____.

  f. We denote the sample size with the symbol _____.

  g. We denote the population median with the symbol _____.

  h. We denote the sample median with the symbol _____.

  i. We denote the population standard deviation with the symbol _____.

  j. We denote the sample standard deviation with the symbol _____.

4. Identify the correct term for the type of bias/effect being defined.

  a. A survey is mailed to 1000 Waco residents. Only 248 of the surveys are returned.

  b. A secondary researcher noticed that subjects changed their answers when the lead researcher raised his eyebrows upon hearing their first response.

  c. The store manager at Big Lots randomly surveyed twenty-five morning shoppers to determine how all of the shoppers felt about the new store hours.

  d. Mr. Hall surveyed fifty high school students asking them if they have ever skipped school.

  e. Students who were viewed cheating by camera did not cheat when people were brought in to monitor their exams.

  f. Although the child only received a sugar pill, she claimed she no longer had a headache.

Introduction to Statistics **39**

5. For the following situations, choose the sampling scheme or method used to obtain data.

   a. To understand the social dynamics of monkeys, a researcher observed a tribe of monkeys in Africa for five days.

   b. From a group of 200 senior adults living at the nursing home, every tenth adult starting with the fourth adult is surveyed to determine their opinion on the new social director.

   c. To poll the nurses working at Scott and White, a researcher used a random number generator to randomly selected one hundred nurses.

   d. A large corporation with 1000 employees divided the employees into ten groups based on the type of work done and sent a questionnaire to twenty randomly selected employees from each of the ten groups to determine what all the employees' opinions are on the new insurance requirements.

   e. To determine what his students feel about his new grading policy, Dr. Smith surveyed each of his students in his class.

   f. To understand what their customers want, Grey Hound Bus Line, randomly selected fifty buses and surveyed every passenger on the bus.

   g. To determine what Baylor students feel about the new proposed mascot emblem used on all official athletic wear, Dr. Smith samples the students in his classes.

6. Suppose data set 1 represents the amount of weight lost in two weeks for twenty randomly selected people following a low carbohydrate diet. Suppose data set 2 represents the amount of weight lost in two weeks for twenty randomly selected people following a low fat diet. Use JMP to answer the following questions.

Data Set 1:	7	9	6	6	5	4	7	6	7	5
	6	6	5	8	4	6	7	6	7	7

Data Set 2:	3	4	2	5	4	5	4	3	4	3
	5	3	3	4	6	3	4	7	6	3

   a. For each data set, find the average amount of weight lost. Is the value calculated μ or $\bar{x}$?

   b. For each data set, find the variance of the data. Is the value calculated σ^2 or s^2?

   c. For each data set, find the coefficient of variation. Which data set has more variability?

   d. Use JMP to create a modified boxplot for each data set on the same axis.

   e. For each data set, find the median.

   f. For each data set, find the Quartiles

   g. Which diet appears to work better? Justify your answer.

7. _____ If the goal is to describe how a particular data value is related to the other data values in the data set, which measure should be calculated?

   a. Measure of Central Tendency.

   b. Measure of Dispersion.

   c. Measure of Location.

8. _____ If the goal is to describe the typical value in a data set, which measure should be calculated?

   a. Measure of Central Tendency.

   b. Measure of Dispersion.

   c. Measure of Location.

9. _____ If the goal is to describe the amount of variability in a data set, which measure should be calculated?

   a. Measure of Central Tendency.

   b. Measure of Dispersion.

   c. Measure of Location.

10. Fill in the blank with either the word *parameter* or *statistic*:

   a. The _____ is a numerical value based on the sample.

   b. The _____ is a numerical value based on the population.

   c. In the field of Statistics, we use the _____ to draw conclusions about the _____.

   d. μ is a _____.

   e. $\bar{x}$ is a _____.

   f. σ is a _____.

   g. s is a _____.

DISCRETE PROBABILITY DISTRIBUTIONS

➡ Section 1: Introduction to Probability:

The theory of probability provides the foundation for statistical inference. I like to think of it as a tool in our tool belt we use for drawing conclusions.

Whenever we draw a conclusion in statistics, we have no idea if we have made the correct decision or an incorrect decision and thus we quantify that uncertainty about our decision using probability. For our purposes, we will only consider the fundamental aspects of probability.

Consider the following: Suppose six out of ten patients suffering from some disease are cured after receiving a particular treatment. Could this have happened even without the treatment or is this evidence that the treatment is effective?

We answer these types of questions using the concepts and laws of probability.

But first, what exactly is probability?

Most statisticians define three different types of probability:

1. *Subjective* probability, which is a measure of personal uncertainty.
2. Probability as *frequency*, which is based on the number of times that a particular event happens during repeated trials of some experiment. This is the type of probability we will use because it is the most widely used in actual practice.
3. *Classical* probability, which uses counting techniques for calculating probability. This is historically associated with games of chance.

➡ Definitions:

- *Probability*: a measure of uncertainty.
  - Probability: is always between 0 and 1, inclusive.
  - If $P(X = a) = 0$, then that implies that "a" _____ occur. "a" is called the Impossible Event
  - If $P(X = a) = 1$, then that implies that "a" and only "a" _____ occur. "a" is called the Certain Event
  - If $P(X = a)$ is "large," then that implies that "a" is   very likely or   not very likely   to occur.
  - If $P(X = a)$ is "small," then that implies that "a" is   very likely or   not very likely   to occur.
  - We have to define what we mean by "small" probability and we will do that in a later chapter.
- *Experiment*: a process in which the outcome can not be predicted ahead of time.

Discrete Probability Distributions 43

- *Sample Space: S* = a collection of all possible outcomes.
  - If describing *S*, are you describing a population or a sample?

  - *N* = the size of the sample space.
- *Event: E* = a subset of the sample space.

  The relationship between Population and Sample versus Sample Space and Event:

Population	Sample Space
Sample	Event

- *Equally Likely Events:* events are equally likely when each outcome in the event has the same chance of occurring.
- *Mutually Exclusive Events:* events that cannot happen at the same time. In other words, there are no elements in common between the events.
- *Exhaustive Events:* when every element in *S* occurs in one of the events. In other words, there are no elements in *S* that are not in one of the events listed.

  For example: Suppose you roll a die. Let E_1 = the die is odd and let E_2 = the die is even.

  Then E_1 = {1, 3, 5} and E_2 = {2, 4, 6}.

  E_1 and E_2 are equally likely events that are also mutually exclusive and exhaustive.

➡ Elementary Properties of Probability:

1. (*The Convexity Law*) Given some process or experiment with *n* mutually exclusive outcomes (called events), $E_1, E_2, ..., E_n$,

   $$0 \le P(E_i) \le 1$$

2. Given some process or experiment with *n* mutually exclusive outcomes,

   $$P(E_1) + P(E_2) + ... + P(E_n) = 1$$

3. For any two events, *A* and *B*

   $$P(A \text{ or } B) = P(A \cup B) = P(A) + P(B) - P(A \cap B)$$

4. (*The Multiplication Rule*) For any two events, *A* and *B*

   $$P(A \text{ and } B) = P(A \cap B) = P(A)P(B \mid A) = P(B)P(A \mid B)$$

5. (*The Multiplication Rule for independent events*)

   $$P(A \text{ and } B) = P(A \cap B) = P(A) \cdot P(B)$$

6. *Probability of Equally Likely Events:* $P(E) = \dfrac{f}{N}$, where *f* is the number of outcomes in event *E*.

- Examples:

Equally likely Events?   Probability?

  - flipping a coin

    $S =$

  - rolling a die

    $S =$

  - shooting a free throw shot

    $S =$

Checking your Understanding: (Answers are found in the Appendix)

1. Determine whether the following probability statements are valid. If invalid, state why.

   a. $P(\mu > 1.2) = 0.0014$

   b. $P(X < 1) = -0.0023$

   c. $P(X > 2) = 1.0011$

Section 1: Homework:

1. Determine whether the following probability statements are valid. If invalid, state why.

   a. $P(X \leq 3.4) = 0.04$

   b. $P(\overline{X} > 2.1) = 0.0004$

   c. $P(\sigma > 1.4) = 0.10$

2. Determine whether the following statements are true or false. If false, state why.

   a. If $P(X = 4) = 0.9999999$, then X is guaranteed to happen.

   b. If $P(X = 2) = 0.00000002$, then this is equivalent to concluding that $P(X = 2) = 0$.

   c. If $P(X = 1) = 0.0000000001$, then X is an improbable event.

3. Suppose I have a die. What is the probability that I roll a 3?

➡ Section 2: Probability Distributions:

Suppose we flip a coin five times. What is S?

Answer:

As we can see, listing all the values of S can be daunting. Fortunately, it is not always necessary. Recall from chapter one that a random variable is a variable that is obtained as a result of a chance process. When calculating probabilities, instead of listing S to describe the population, we will use random variable notation to describe the population. Here is another definition of random variable:

Discrete Probability Distributions **45**

Random variable – a rule that assigns a numerical quantity to each outcome in the sample space. We generally denote random variables with X, Y, or Z.

- Example: Suppose we flip a coin three times and let X = the number of heads.

    S =

    X =

- Random variables are either discrete or continuous.
- In inferential statistics, we are trying to draw conclusions about the population based on what we have seen from our sample.
- We represent our uncertainty about our conclusions with probability. Thus, to every value of a variable, we can attach a probability representing our uncertainty about that value.

    This is the *probability distribution* of the random variable.

Definition:

- The *probability distribution* describes the population of data values. In other words, the probability distribution describes *all* the possible values that can be attained by the random variable.

    - Is the probability distribution describing the population or a sample?

    Answer:

The probability distribution depends on the type of random variable obtained:

- If the random variable is continuous, we describe its distribution with a continuous curve, called a *probability density function* (PDF).
- If the random variable is discrete, we describe its distribution with a probability mass function (PMF).

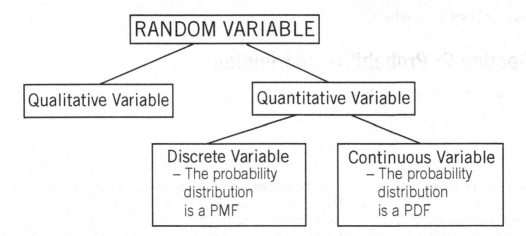

46   Introduction to Statistics

Section 3: Discrete Probability Distributions:

Probability Mass Function, <u>PMF</u>: - The probability distribution of a discrete random variable is a table, graph, formula, or any device used to specify all possible values of a discrete random variable along with their respective probabilities.

For example:

X	2	3	4	5	6
f(x)	0.1	0.4	0.1	0.3	0.1

Notation: $f(x) = P(X = x)$

$f(3) =$

The PMF gives probability _____ _____ .

The graph of a PMF is called a *Probability Histogram*.

To make a probability histogram:

1. The horizontal axis represents the random variable.
2. The vertical axis represents the probability.
3. The probability is represented by a bar with width one that is centered about the random variable and has height equal to the probability.

Area of the bar = _____ .

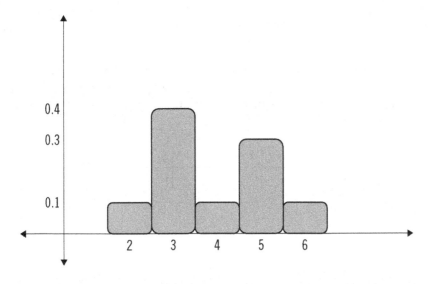

- Example 1: Consider flipping a coin three times. Let X = # of heads.
  - Find the PMF:
    Answer:

- Graph the PMF:

  Answer:

- Example 2: Ava has a four sided green die with the numbers 1, 2, 3, 4 on the sides. (One side has a 1, one side has a 2, etc.) Ali has a six sided pink die with the numbers 1, 2, 3, 4, 5, 6. Suppose Ava and Ali each roll their die one time. Let X = sum of the green and pink dice. Find the PMF for X.

  Answer:

- Properties: For a function to be a PMF, the following must be true.

  1. $0 \le f(x) \le 1$

  2. $f(x) = 0$ for all x not in S

  3. $\sum_{x \in S} f(x) = 1$

- Cumulative Distribution Function, _CDF_ - the cumulative distribution function of a discrete random variable is a table, graph, formula, or any device used to specify all possible values of a discrete random variable along with their respective cumulative probabilities.

  For example:

X	2	3	4	5	6
$F(x)$	0.1	0.5	0.6	0.9	1.0

  Notation: $F(x) = P(X \le x)$

  $F(3)=$

The CDF gives _____ _____.

48   Introduction to Statistics

We find the CDF, by simply summing up the PMF. Thus, $F(x) = P(X \leq x) = \Sigma f(x)$.

The graph of the CDF is a step function.

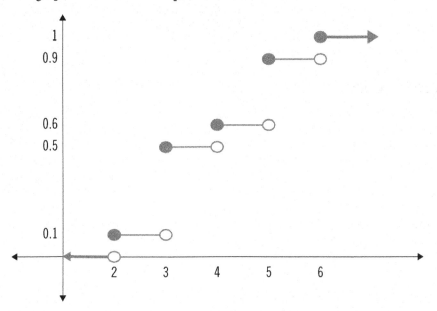

To create the step function:

1. The horizontal axis represents the random variables.
2. The vertical axis represents the cumulative probability.
3. Open circles represent when not including the random variable and closed circles are used to represent when we are including the random variable. Arrows are used to represent the cumulative probability at that value of the random variable.

- Example 3: Find the CDF using the PMF found in Example 1:
  Answer:

- Graph the CDF:
  Answer:

Discrete Probability Distributions **49**

- Example 4: Find $F(x)$ for our example 2:

  Answer:

- Properties of $F(x)$, where $F(x)$ is a CDF:

  1. $0 \leq F(x) \leq 1$

  2. $\lim_{x \to -\infty} F(x) = 0$ and $\lim_{x \to +\infty} F(x) = 1$

  3. If $a < b$, $F(a) \leq F(b)$

  4. $F(x)$ is right continuous

- <u>Using the PMF or CDF to calculate probabilities:</u>

  Recall, $f(x) =$

  Whereas

  $F(x) =$

To calculate a probability using the PMF, $f(x)$, determine which random variables satisfy the range and sum the corresponding probabilities.

- Example 5: Suppose you are given the following PMF for the RV X, where X represents the number of successes.

X	0	1	2	3	4	5
$f(x)$	0.2	0.1	0.3	0.1	0.1	0.2

  a. Find $P(2 < X < 5)$

  Answer:

  b. Find probability of having at least 4 successes:

  Answer:

Since the CDF is defined as, $F(x) = P(X \leq x)$, we calculate the probability of being less than or equal to or strictly less than by evaluating the CDF function at the appropriate value of x. If we want to calculate the probability of being greater than or greater than or equal to, we use the complement rule. If we want to find the probability of being between two values, we subtract the two CDF functions evaluated at the appropriate values of x. (To determine the appropriate value of x, draw a snapshot of the histogram.)

This gives us the following formulas:

$P\left(X \genfrac{}{}{0pt}{}{<}{\leq} a\right) = $ _____

$P\left(X \genfrac{}{}{0pt}{}{>}{\geq} a\right) = $ _____

$P\left(a \genfrac{}{}{0pt}{}{<}{\leq} X \genfrac{}{}{0pt}{}{<}{\leq} b\right) = $ _____

- Example 6: Once again suppose you have the PMF from Example 5, where X represents the number of successes.

X	0	1	2	3	4	5
f(x)	0.2	0.1	0.3	0.1	0.1	0.2

   a. Find the CDF.

   Answer:

   b. Find probability of having at most 2 successes.

   Answer:

Discrete Probability Distributions **51**

c. Find probability of having between 2 and 4 successes, inclusive.

Answer:

d. Find probability of having between 1 and 4 successes. (Note that between always implies "exclusive". If you want the endpoints included, you must say "inclusive".)

Answer:

Mean and Variance of a Discrete Probability Distribution:

- When describing a sample, one should always include three things: a "picture" of the data using a graph, a measure of central tendency, and a measure of dispersion. This is true when describing a population. The probability histogram gives us a "picture" of the data (or the step graph of the CDF). We also would like to know a measure of central tendency and a measure of dispersion. The measure of central tendency we will discuss is the population mean, which is also known as the expected value. The measure of dispersion we will discuss is the variance or standard deviation.

The Expected Value: It is often of interest to find the long-run average of a particular distribution.

- This long-run average is known as the expected value or mean value of the random variable

  - Formally, the expected value of a random variable X is the mean value of the variable in the sample space or population of possible outcomes. Think of it as the mean of what we would get if we did the experiment over and over and over again...what we would get in the long run... what we *expect* to get in the long run.

52   Introduction to Statistics

- Calculating Expected Value

  If X is a random variable with possible values x_1, x_2, x_3,... occurring with probabilities $f(x_1), f(x_2), f(x_3),$ ...then

  $$E(X) = \mu = \sum x_i f(x_i)$$

- Variance and Standard Deviation for a Discrete Random Variable

  - Recall that variance and standard deviation describe how much variability there is in the data by comparing how spread out the values are about the mean. This gives us an idea of how far from the mean we expect values to fall for data from a given distribution.

  - Calculating variance: If X is a random variable with possible values x_1, x_2, x_3,... occurring with probabilities $f(x_1), f(x_2), f(x_3),$ ...and with expected value $E(X) = \mu$ then

    $$\text{Variance of } X = V(X) = \sigma^2 = \Sigma(x_i - \mu)^2 f(x_i) = \Sigma x^2 f(x) - \mu^2$$

  - Standard deviation of X: $\sigma = \sqrt{\Sigma x^2 f(x) - \mu^2}$

- Example 7: Suppose you are going to gamble on a roulette wheel, where you pay $1.00 to place your bet. (The roulette wheel has thirty-eight equal spaces, eighteen red, eighteen black, and two green.) If the ball falls on the black eight, you win $3.00. Let $X =$ the *net* winnings.

  Find and interpret the expected value:

  Find the variance:

Discrete Probability Distributions **53**

Interpretation:

Checking your Understanding: (Answers are found in the Appendix)

1. Suppose the cumulative probability distribution for the random variable X is as follows:

X	0	0.5	1	1.5	2	4
$F(x)$	0.47	0.62	0.73	a	0.98	b

   a. If the $P(X = 1.5) = 0.08$ what is the value of a?

   b. What is the value of b?

   c. Find $P(X = 2.5)$

   d. Find the probability that X is at least 1.5

   e. Find $P(0 < X \leq 2)$

2. Suppose you play a dice game where if you roll a composite number, you win $5.00. If you roll a one, you win $10. If you roll a prime number, you lose $8. You must pay $2.00 to play. Let $X =$ net winnings.

   a. Find the PMF.

   b. Find the expected value.

   c. Interpret the expected value.

Section 3: Homework:

1. _____ Choose the choice that gives the PMF notation.

   a. $f(x) = P(X = x)$
   b. $f(x) = P(X \leq x)$
   c. $f(x) = P(X < x)$
   d. $f(x) = P(X \geq x)$
   e. $f(x) = P(X > x)$

   f. $F(x) = P(X = x)$
   g. $F(x) = P(X \leq x)$
   h. $F(x) = P(X < x)$
   i. $F(x) = P(X \geq x)$
   j. $F(x) = P(X > x)$

2. _____ Choose the choice that gives the CDF notation.

   a. $f(x) = P(X = x)$
   b. $f(x) = P(X \leq x)$
   c. $f(x) = P(X < x)$
   d. $f(x) = P(X \geq x)$
   e. $f(x) = P(X > x)$

   f. $F(x) = P(X = x)$
   g. $F(x) = P(X \leq x)$
   h. $F(x) = P(X < x)$
   i. $F(x) = P(X \geq x)$
   j. $F(x) = P(X > x)$

54   Introduction to Statistics

3. For each table below, determine whether the table represents a PMF. If not, state why not.

a.

x	-3	-1	0	1	3
f(x)	0.24	0.19	0.45	0.30	0.02

b.

x	0	4	8	12	16
f(x)	0.24	0.19	0.15	0.30	0.02

c.

x	1	6	11	16	21
f(x)	0.14	0.29	0.05	0.30	0.22

4. For each table below, determine whether the table represents a CDF. If not, state why not.

a.

x	−3	−1	0	1	3
F(x)	−0.24	0.19	0.45	0.55	1

b.

x	0	4	8	12	16
F(x)	0.20	0.39	0.35	0.45	1

c.

x	1	6	11	16	21
F(x)	0.14	0.29	0.35	0.59	1

5. Consider the experiment of rolling a die. Find the PMF.

6. Consider the experiment of rolling a die. Find the CDF.

7. A college graduate statistics class has eighteen students. The ages of these students are as follows: one student is 18 years old, two are 20, nine are 22, three are 24, two are 29, and one is 40. Let X = age of any student (randomly selected). Find the probability mass function for X.

8. Five thousand instant lottery tickets were sold. One ticket has a face value of $10,000, five tickets have face values of $1000 each, twenty tickets are worth $500 each, thirty tickets have face value of $100 each, and 500 are worth $1.00 each. The rest are losers. Let X = the value of a ticket that you buy. Find the probability mass function for X.

9. Suppose the probability distribution for X = number of jobs held during the past year for work study students at a university is as follows:

X	0	1	2	3	4	5
f(x)	0.12	a	0.28	0.18	0.06	0.02

a. Find a.

b. Find the probability that a randomly selected student held three or more jobs during the past year.

c. Find $P(1 < X < 4)$.

d. Find the cumulative distribution function.

e. Find the $E(X)$.

f. How many jobs would you expect a randomly selected student to hold in the past year?

Discrete Probability Distributions 55

10. Use the CDF given below to answer the following questions.

X	0	1	2	3	4	5	6	7	8	9	10
F(x)	0.24	0.29	0.35	0.38	0.40	0.45	0.56	0.62	0.88	0.92	a

a. Find a

b. Find the probability of having at most 5.

c. Find the probability of having at least 8.

d. Find the probability of being between 2 and 8.

11. Since Carl travels quite a bit for his work, he rents his car. So he is debating whether or not he should purchase car rental insurance. The car rental insurance is $40 a year. He thinks that the probability that he will need to make only one claim a year on his rental car is 0.1, the probability that he has to make only two claims per year is 0.05, and the probability that he will have to make more than two claims a year is zero. It will cost Carl $120 each time if his rental car has to be "fixed" if he doesn't have insurance but will cost nothing if he does have insurance. Let $X =$ Carl's cost next year for "fixing" his rental car and/or car rental insurance.

a. If Carl buys the car rental insurance, what is the PMF? (Hint: X has only one possible value.)

b. If Carl doesn't buy the insurance, what is the PMF? (Hint: X has three possible values.)

c. For parts a and b, find the E(X).

d. Using these two values found in part c, should Carl buy the insurance? Explain why or why not.

12. Use the following PMF to answer the question below.

x	0	2	4	6	8	10
f(x)	0.23	0.12	c	0.10	0.36	0.05

a. Find the probability that X equals 4.

b. Find the probability that X equals 2.

c. Find the probability that X is at most 2.

d. Find the probability that X is at least 8.

e. Find the probability that X is between 2 and 8.

f. Find the probability that X is odd.

➡ Section 4: The Binomial Distribution:

When finding the probability mass function, finding the probability can sometimes be challenging. Some experiments have specific characteristics. These specific characteristics can be modeled by a specific, well-defined discrete probability distribution. The benefit to this is that if we recognize the characteristics of a well defined distribution, then there will be a known formula or table that we can use for calculating the probabilities. One such well known discrete probability distribution is the Binomial distribution.

- *Bernoulli Trial* – when a random process or experiment results in one of only two mutually exclusive outcomes, the trial is called a *Bernoulli Trial*.

56 Introduction to Statistics

- *Bernoulli Process* – a sequence of Bernoulli trials such that:

  1. Each trial has only two mutually exclusive outcomes. One is noted as a success, the other as a failure.

  2. The probability of success, p, remains constant from trial to trial. The probability of failure is denoted as q.

  3. Trials are independent. The outcome of one trial is not affected by the outcome of any other trial.

- *Binomial Distribution* – if our random variable, X, is defined to count the number of successes and our experiment is a finite Bernoulli process, then the resulting distribution is a *Binomial distribution*.

- Thus, for X to be a Binomial RV, there are two things we must verify:

  1. X must count the # of successes

  2. The experiment must be a finite Bernoulli process.

- If the random variable X is a Binomial random variable, we denote it as $X \sim Bin(n, p)$.

  - Example 1: Are the following random variables Binomial random variables?

    a. Sixty percent of all commercial bus drivers are over thirty years of age. Out of a random sample of twenty-five drivers, let the random variable X be the number of bus drivers who are over thirty years of age.

    Answer:

      1. Is the RV counting the number of successes?

      2. Is it a finite Bernoulli Process?
         i. Finite number of trials?

         ii. Only two outcomes?

         iii. Constant probability?

         iv. Independent trials?

    b. Suppose a clothing salesperson makes sales to 35% of her customers. One day she counted her customers until she made three sales. Let X be the number of customers until her third sale.

    Answer:

      1. Is the RV counting the number of successes?

Discrete Probability Distributions **57**

2. Is it a finite Bernoulli Process?

    i. Finite number of trials?

    ii. Only two outcomes?

    iii. Constant probability?

    iv. Independent trials?

c. A department of a large corporation has ten employees, seven women and three men. Four employees are to be selected to form a committee. Let the random variable X be the number of women on the committee.

Answer:

    1. Is the RV counting the number of successes?

    2. Is it a finite Bernoulli Process?

        i. Finite number of trials?

        ii. Only two outcomes?

        iii. Constant probability?

        iv. Independent trials?

- The Binomial Probability Mass Function:
  The PMF for the Binomial distribution is

$$f(x) = P(X = x) = \binom{n}{x} p^x q^{n-x} \quad X = 0, 1, ..., n$$

where n = the number of trials

    p = the probability of success

    q = the probability of failure.

- If we know X is a Binomial random variable, then we know the summary parameters as well:

  The mean: $\mu = n \cdot p$

  The variance: $\sigma^2 = n \cdot p \cdot q$

  The standard deviation: $\sigma = \sqrt{n \cdot p \cdot q}$

- Example 2: Suppose $X \sim Bin(10, 0.6)$

  JMP Output:

X	PMF	CDF
0	0.0001048576	0.0001048576
1	0.001572864	0.0016777216
2	0.010616832	0.0122945536
3	0.042467328	0.0547618816
4	0.111476736	0.1662386176
5	0.2006581248	0.3668967424
6	0.250822656	0.6177193984
7	0.214990848	0.8327102464
8	0.120932352	0.9536425984
9	0.040310784	0.9939533824
10	0.0060466176	1

- Find P(X = 2), using the PMF:

  Answer:

- Find P(X = 3), using the CDF:

  Answer:

- Find P($X \geq 2$) using the PMF:

  Answer:

Discrete Probability Distributions **59**

- Find $P(X \geq 3)$ using the CDF:

  Answer:

- Example 3: Seventy percent of the teachers in a large town are opposed to a proposed educational amendment. Suppose a researcher is interested in calculating probabilities about the number of teachers opposed and randomly samples twenty teachers.

  a. What is the random variable, X?

  b. What is the distribution of X?

  c. Find the probability that more than eighteen teachers are opposed.

d. Find the probability that fewer than two teachers are opposed.

e. Find the probability that between ten and fifteen teachers are opposed.

JMP output:

X	PMF	CDF
0	3.486784e-11	3.486784e-11
1	1.6271661e-9	1.6620339e-9
2	3.6068848e-8	3.7730881e-8
3	5.0496387e-7	5.4269475e-7
4	5.0075583e-6	5.5502531e-6
5	0.0000373898	0.00004294
6	0.000218107	0.000261047
7	0.0010178326	0.0012788796
8	0.0038592819	0.0051381615
9	0.0120066549	0.0171448164
10	0.0308170809	0.0479618973
11	0.0653695655	0.1133314629
12	0.1143967397	0.2277282026
13	0.1642619852	0.3919901878
14	0.1916389828	0.5836291706
15	0.1788630506	0.7624922211
16	0.1304209744	0.8929131955
17	0.0716036722	0.9645168677
18	0.0278458725	0.9923627402
19	0.0068393371	0.9992020773
20	0.0007979227	1

Discrete Probability Distributions **61**

JMP Steps for finding the Binomial PMF and/or CDF:

Step 1: Create a column representing the values of X.

- Under the *Col* menu, choose *new column*, name it X.
- Under missing/empty, choose *sequence data*.
- Put 0 for the *From* and the value of *n* for the *To*.
- Choose *ok*.

Step 2: Create the PMF or CDF.

- Under the *Col* menu, choose *new column,* name it PMF or CDF
- Under *Column Properties*, choose *formula*
- On the left side of the menu, choose *Discrete Probability* and then choose *Binomial Probability* for the PMF or *Binomial Distribution* for the CDF
- Fill in the values of *n* and *p*
- To input *k*, click on the column representing *X* and click on *k*.

Checking your Understanding: (Answers are found in the Appendix)

Montana Table Company produces custom-made furniture for restaurants, diners and hotels. All of their furniture is handmade with it being known that 15 percent of the tables have some sort of flaw. Smitty's Steak House, a large restaurant, orders 15 new tables from Montana Table Company.

a. What is the probability that at least three will be flawed?

b. How many tables ordered by Smitty's Steak House would you expect to have a flaw?

Section 4: Homework:

1. Suppose Amy works in the advertisement sales office of the local cable station. Amy successfully sells an advertisement to 65% of her contacts. Suppose Amy has enough time to contact twenty people per day. Let *X* = the length of time between successful sales of advertisement. Is the random variable *X* a binomial random variable? If so, give the parameters. If not, explain why not.

2. Flying Express Airlines flight 82 from Dallas to Miami is on time 85% of the time, according to Flying Express Airlines. Suppose sixty-five flights are randomly selected.

   a. What does the random variable *X* represent?

   b. What is the distribution of the random variable *X*?

   c. How many flights from Dallas to Miami would you expect to arrive on time?

3. Suppose it is known that a person has a 5% chance of obtaining a winning lottery ticket in a state lottery. Suppose she plays the game fifteen times.

   a. What does the random variable *X* represent?

   b. What is the distribution of the random variable *X*?

   c. Use JMP to find the PMF.

   d. Use JMP to find the CDF.

62 Introduction to Statistics

e. Use the PMF to find the probability that the player wins at most two times.

f. Use the CDF to find the probability that the player wins at least five times. (For parts e and f, you must show your work to verify that you used the requested method)

4. According to *Bicycle Weekly News*, Davis, California is one of the top ten bicycle friendly towns in America. In the town of Davis 64% of the population owns a bicycle. Suppose a random sample of thirty-five residents of Davis is selected and the town mayor is interested in calculating the probability of owning a bike.

   a. What is the random variable X?

   b. What is the distribution of X?

   c. Use JMP to find the CDF. Use the CDF to calculate the probability that between twenty and thirty residents sampled own a bicycle. (You must show your work to verify you used the requested method)

   d. How many of the residents sampled would you expect to see own a bicycle?

5. Suppose it is known that 86% of Americans make New Year's resolutions. Further suppose that Outliers Express has twenty-eight employees and the president of the company is interested in the number of his employees who made a New Year's resolution.

   a. What is the random variable X?

   b. What is the distribution for X?

   c. Use JMP to find the PMF. Use the PMF to calculate the probability that more than twenty-five employees of Outliers Express made a New Year's resolution. (You must show your work to verify that you used the requested method.)

   d. Find μ.

   e. Find σ.

➡ Section 5: Discrete Probability Distributions Overview:

Probability Distributions—

Describes _____ possible values that can be obtained by the random variable.

X is Discrete

PMF: $f(x) =$

Discrete Probability Distributions **63**

CDF: $F(x) =$

Suppose X is a discrete RV with possible values, $X = 0, 1, 2, ..., 10$.
Find the following probabilities using $f(x)$ and $F(x)$.

$P(X \le 2)$:

$P(X > 2)$:

$P(2 < X < 5)$:

Binomial: $X \sim \text{Bin}(n,p)$:

➡ Chapter 2 Review Homework:

1. Let $f(x)$ given below represent the PMF of the random variable X.

$$f(x) = \begin{cases} 1/5 & \text{if } X = -2, -1, 0, 1, 2 \\ 0 & \text{elsewhere} \end{cases}$$

   a. Re-write the above PMF as a table.

   b. Graph $f(x)$.

   c. Find the cumulative distribution function, $F(x)$ of X.

   d. Graph $F(x)$.

2. Suppose Ana has a pair of dice (the traditional six-sided kind). Let X = the difference of the largest minus the smallest number showing on the dice. Find the PMF for X.

3. Suppose in a game of cards, if an Ace card is drawn, the player will win $5.00. The player must pay $2.00 to play. Suppose the game is played one time. Let X = the players net winnings.

   a. Find the PMF.

   b. Find and interpret the $E(X)$.

   c. Find $V(X)$.

   d. Find σ

4. Suppose James has a box with eight balls in it; four red, four green. Suppose James is going to randomly draw three balls without replacement from the box. Let the random variable X be the number of red balls drawn. Is the RV a Binomial RV?

5. Of all the registered automobiles in Texas, 58% are sport utility vehicles. Suppose Carol's neighborhood has 20 vehicles registered.

   a. What is the probability that a majority of them are sport utility vehicles?

   b. How many sport utility vehicles would you expect to find in Carol's neighborhood?

6. Let $f(x) = \dfrac{5!}{x!(5-x)!} \left(\dfrac{1}{3}\right)^x \left(\dfrac{2}{3}\right)^{5-x}$ $X = 0, 1, 2, 3, 4, 5$

   Verify that $f(x)$ is a probability mass function (PMF). Hint: there are two things to verify.

Discrete Probability Distributions **65**

CONTINUOUS PROBABILITY DISTRIBUTIONS

Recall, Probability Distributions describe _____ possible values that can be obtained by the RV.

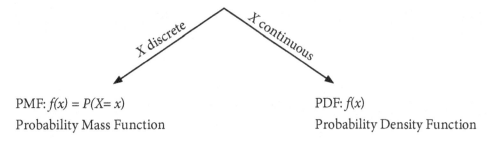

PMF: $f(x) = P(X= x)$
Probability Mass Function

PDF: $f(x)$
Probability Density Function

▶ Section 1: The Probability Density Function, PDF:

The Probability Density Function, PDF, for a continuous random variable is the curve $f(x)$ that is attained from smoothing out the histogram that is obtained through repeated sampling. For us, $f(x)$ will be given.

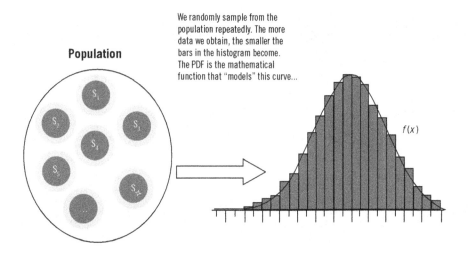

- Properties:
  1. Probability is equal to area under the PDF curve.
  2. $P(a < X < b)$ = area under the PDF curve from a to b.
  3. The total area under the curve equals 1.
  4. $f(x) \geq 0$ for all X in the range
  5. $P(X = a) = 0$

     Why?

     Answer:

Example: Suppose you have the following PDF for the continuous RV, X.

Find $P(X > 3)$

Checking your Understanding (Answers are found in the Appendix)

1. For a continuous RV, $P(a < X < b) =$ _____.

2. Suppose you have the following PDF for the continuous RV, X. Find $P(X > 4)$

Section 1: Homework:

1. Consider the following probability distribution for the continuous random variable, X.

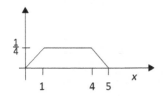

   a. Find $P(2 < X < 4)$
   b. Find $P(4 < X < 5)$
   c. Find $P(X < 4)$

2. The elementary physical education teacher recorded the length of time, X, in minutes, required for fourth graders to run one mile. The probability distribution is as follows:

   In parts (a) – (d), find the percentage of fourth graders that ran the mile in the specified time. Write your answer accurately to two decimal places.

   a. More than 15 minutes
   b. Less than 8 minutes
   c. Between 8 and 12 minutes
   d. More than 14 minutes

Continuous Probability Distributions 69

Section 2: The Cumulative Distribution Function, CDF:

The Cumulative Distribution function, $F(x)$: The Cumulative Distribution Function gives cumulative probability. Since for a continuous distribution, probability is equal to area under the curve, the formula for finding $F(x)$ is:

$$F(x) = P(X \leq x) = \int_{-\infty}^{x} f(y)dy$$

- Example:

- Find $P(X \leq 2)$ using the PDF:

   Answer:

- Find $P(X \leq 2)$ using the CDF:

   Answer:

Example 1: Given the CDF, $F(x) = \begin{cases} 0 & X < -2 \\ \dfrac{x^3 + 6x + 20}{40} & -2 \leq X \leq 2 \\ 1 & X > 2 \end{cases}$

a. Find $P(X \leq 1)$

b. Find $P(1 < X < 3)$

c. Find $P(X > 1.5)$

Checking your Understanding (Answers are found in the Appendix)

Suppose you have the following CDF for the random variable, X.

$$F(x) = \begin{cases} 0 & X < 1 \\ \frac{3}{50}\left[\frac{x^3}{3} + x^2 - \frac{4}{3}\right] & 1 \le X \le 3 \\ 1 & X > 3 \end{cases}$$

    a. Find $P(X \le 1)$

    b. Find $P(X > 2)$

Section 2: Homework:

1. Given the following CDF, $F(x) = \begin{cases} 0 & X < -1 \\ \frac{x^3+1}{2} & -1 \le X \le 1 \\ 1 & X > 1 \end{cases}$

    a. Find $P(X \le \frac{1}{2})$

    b. Find $P(-2 < X < 0)$

    c. Find $P(-\frac{1}{2} < X < \frac{1}{2})$

    d. Find $P(X > \frac{1}{4})$

    e. Find $P(X = \frac{1}{2})$

➡ Section 3: The Uniform Distribution:

Continuous random variables that have equally likely outcomes over their range of possible values have a Uniform Probability Distribution.

Since all values of a Uniform RV have the same probability of occurring, the height of each bar in the histogram is the same. Thus the bars smooth out into a horizontal line, giving us the following graph:

- The graph of $f(x)$:

- Notation: $X \sim Unif(c, d)$

- The PDF

  To find the PDF, recall the total area under the curve must equal 1.

  Thus, we know that   base*height $= 1$.
                           $(d - c)*h = 1$.

  Solving for h, we have $h = \dfrac{1}{d - c}$

  Hence, the PDF is: $f(x) = \dfrac{1}{d - c}$ $c \le X \le d$

72   Introduction to Statistics

- The mean: $\mu = \dfrac{c+d}{2}$
- The standard deviation: $\sigma = \dfrac{d-c}{\sqrt{12}}$

- Recall, Probability = Area Under the Curve.

  Thus, $P(a < X < b)$ = area under the curve from a to b

  =

Example: Suppose $X \sim Unif(2, 8)$.

a. Find $f(x)$

b. Graph $f(x)$

c. Find $P(X > 3)$

Checking your Understanding: (Answers are found in the Appendix)

Sue's Diner orders their chairs from Montana Table Company. When Sue calls Montana Table Company to place an order, she spends 5 to 30 minutes waiting on hold where the probability of waiting on hold is constant for all minutes between 5 and 30. Suppose Sue is interested in calculating the probabilities about the length of time she waits on hold.

a. Find the PDF for X.

b. Find the probability that Sue spends between 10 and 20 minutes on hold.

Continuous Probability Distributions **73**

Section 3: Homework:

1. Suppose the probability of a continuous random variable X, occurring over the range from two to seven, is constant.
   a. Draw a picture of the PDF.
   b. Find the PDF function, $f(x)$.
   c. Find $P(3 < X < 6)$

2. Suppose $X \sim Unif(20, 60)$.
   a. Find the probability that X is at most 25.
   b. Find the probability that X is at least 40.

3. Suppose Tommy is a delivery driver for Jed's Crispy Pizza Junction. It is known that the time it takes to deliver a pizza has a uniform distribution from twenty minutes to sixty minutes.
   a. What is the probability that the time to deliver a pizza is less than thirty-five minutes?
   b. If the time to deliver the pizza is longer than fifty minutes, the customer will receive $5.00 off the price of the pizza. What is the probability that the customer will receive $5.00 off the price?

➡ Section 4: The Normal Distribution:

- Characteristics of the Normal Distribution:
  1. The PDF is $f(x) = \dfrac{1}{\sqrt{2\pi}\,\sigma} e^{\dfrac{-(x-\mu)^2}{2\sigma^2}} \quad -\infty < X < +\infty$
  2. Symmetric bell-shaped curve
  3. Each curve is determined by the parameters μ and σ.
     μ is known as the location parameter; σ is known as the shape parameter.

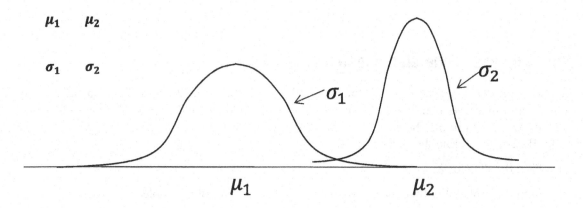

74   Introduction to Statistics

4. Notation: $X \sim N(\mu, \sigma)$

5. The Standard Normal Curve, $N(0, 1)$. Each normal curve can be standardized by using the formula:
$$Z = \frac{x - \mu}{\sigma}$$

The z-score, Z, is the standardized score of the random variable X and describes for us how many standard deviations the random variable X is above or below the mean.

Example: Suppose the random variable X = test scores $\sim N(80, 5)$

What is the z-score for $X = 85$?

Answer:

What is the z-score for $X = 90$?

Answer:

What is the z-score for $X = 75$?

Answer:

What is the z-score for $X = 70$?

Answer:

What is the z-score for $X = 80$?

Answer:

The z-score is just another way for representing the random variable X. In other words, I can tell you that you made a 95 on the test or that you made a standardized score of 3 and I am telling you the exact same thing.

$P(75 < X < 85) =$

Continuous Probability Distributions 75

The Normal Curve and the Emperical Rule:

$P(-1 < Z < 1) \approx$ _____

$P(-2 < Z < 2) \approx$ _____

$P(-3 < Z < 3) \approx$ _____

- Example 1: Suppose $X \sim N(11, 2)$.

    a. Find $P(X < 12)$

    Answer:

    b. Find $P(6 < X < 13)$

    Answer:

c. Find $P(X > 7.62)$

   Answer:

- A special notation with z-scores:

   Z_α: This notation represents the specific z-score that has area to the right equal to α.

   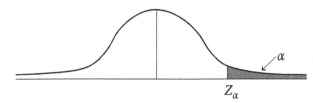

   Thus, if asked to find Z_α, find the z-score that has AREA to the RIGHT equal to α.

   Example 2:

   a. Find $Z_{0.05}$:

   b. Find $Z_{0.95}$

Continuous Probability Distributions 77

- Example 3: Suppose again that $X \sim N(11, 2)$.

   a. Find x such that $P(X < x) = .025$

   Answer:

   b. Find x such that $P(X > x) = .10$

   Answer:

   c. Find x such that $P(X > x) = .95$

   Answer:

   d. Find x such that $P(X < x) = .85$

   Answer:

- Example 4: The height of adult women in France is a normally distributed random variable with a mean of 64 inches and a standard deviation of 2.4 inches. Find the probability that a randomly chosen woman is:

a. Less than 62 inches tall.

Answer:

b. More than 67 inches tall.

Answer:

c. Between 61 and 66 inches tall.

Answer:

d. If only 5% of the population of women in France is taller than Marie, what is her height in inches?

Answer:

- Example 5: The Triglyceride levels in healthy American adults are normally distributed with a mean of 134 mg/dL and a standard deviation of 14 mg/dL.

  a. If a healthy adult is randomly selected, what is the probability that their Triglyceride level is between 150 and 180?

  Answer:

  b. Physicians are told to recommend exercise and dietary changes if one's Triglyceride level is in the upper 3%. What is the cutoff for healthy adults?

  Answer:

Checking your Understanding: (Answers are found in the Appendix)

1. Find $Z_{0.90}$

2. Find $Z_{0.30}$

3. The heights of cottonwood trees in Texas are normally distributed with a mean of 90 feet and standard deviation of 10 feet.

   a. Find the probability that a randomly selected cottonwood tree is shorter than 75 feet.

   b. What is the probability that a randomly selected cottonwood tree is between 70 and 105 feet tall?

   c. Find the probability that a randomly selected cottonwood tree is taller than 97 feet.

   d. Find the height such that only 10% of trees are taller than that.

80   Introduction to Statistics

Section 4: Homework:

1. Suppose Z represents the standard normal random variable. If a value is selected at random from the standard normal distribution, find the probability that Z is:

   a. Less than 0

   b. Between –2.35 and 1.65

   c. Greater than 1.86

   d. Greater than –2.15 and less than 1.67

   e. Less than –1.92 or greater than 2.41

2. Find the following z-scores:

   a. $Z_{0.10}$

   b. $Z_{0.975}$

   c. $Z_{0.005}$

   d. $Z_{0.15}$

   e. $Z_{0.80}$

   f. $Z_{0.30}$

3. Suppose $X \sim N(10, 2)$

   a. Find the probability that X is at least 8.5.

   b. Find the probability that X is at most 7.

   c. Find the probability that X is between 6 and 14.

   d. Find x such that $P(X < x) = 0.99$.

   e. Find x such that $P(X > x) = 0.30$.

4. Suppose the incubation period for land turtle eggs is normally distributed with a mean of sixty-seven days and standard deviation 7.4 days.

   a. Find the probability that a land turtle egg will take at least eighty days to hatch.

   b. Find the probability that a land turtle egg will take at most fifty-five days to hatch.

   c. Find the probability that a land turtle egg will take between fifty and eighty-five days to hatch.

   d. Find the length of time such that 85% of the land turtle eggs will have hatched by then.

   e. Suppose a veterinarian becomes concerned about the health of the land turtle if it takes longer than ninety days to hatch. What is the probability of this happening if the turtle is healthy? Is it likely for this to happen?

   f. Find and interpret the 90th percentile.

Continuous Probability Distributions **81**

5. Suppose the length of time of a whale's song is normally distributed with mean 12.5 minutes and standard deviation 2.6 minutes.

   a. Find the probability that a whale's song will last more than ten minutes.

   b. Find the probability that a whale's song will last between eight and sixteen minutes.

   c. Find the probability that a whale's song lasts less than five minutes or longer than twenty minutes.

   d. Find the length of time such that only 5% of whale's songs last longer than that length of time.

   e. Find the length of time such that only 10% of whale's songs are shorter than that length of time.

6. The Downtown Marcela Street subway is docked at the station for passengers to enter all morning, and the time that it first departs from the station is typically some amount of minutes after 9:00. Let X be the number of minutes after nine o'clock that the Downtown Marcela Street subway leaves the station. Assume that the distribution of times is approximately normal with mean fifteen minutes and standard deviation five minutes.

   a. If a person gets to the subway station at 9:10, what is the probability that the person has missed the Downtown Marcela Street subway?

   b. If a person is willing to risk a 15% chance of not making the Downtown Marcela Street subway, what is the maximum number of minutes after nine o'clock that the person can reach the station?

   c. What time should the person reach the station to have a 50% chance of catching the Downtown Marcela Street subway?

➡ Section 5: Continuous Probability Distribution Overview:

Probability Distributions— Describes _____ possible values that can be obtained by the random variable.

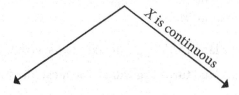

PDF: $f(x) =$

probability = _____ under the PDF, $f(x)$, curve.

Total area = _____ .

$P(X = a)$ = _____ .

CDF: $F(x)$ =

Probability is found by function evaluation.

$P(X \leq a) = P(X < a)$ = _____ .

$P(X \geq a) = P(X > a)$ = _____ .

$P(a \leq X \leq b) = P(a < X < b)$ = _____ .

Uniform: all RV's have the _____ probability of occurring.

    f(x):

    To find the probability of a Uniform RV occurring, you simply find the _____ of the

    _____ .

Normal: $X \sim N(\mu, \sigma)$

    Find probabilities by finding the *z*-score $=$

    $P(Z < a) =$

$P(Z > a) =$

$P(a < Z < b) =$

Continuous Probability Distributions   **85**

Probability Distributions – describe _____ possible values that can be obtained by the RV.

continuous

- **PDF:** $f(x)$: The mathematical function that is found by smoothing out the histogram that is obtained through repeated sampling.

  Probability = _____ under the PDF curve

- **CDF:** $F(x)$ =

  Probability is found by function evaluation.

  i.e. $P(X \leq a) = P(X < a) =$

  or $P(X \geq a) = P(X > a) =$

  or $P(a < X < b) = P(a \leq X \leq b) =$

Uniform: all RV's have same probability of occurring.

Find probability by finding _____ of rectangle

- $N(\mu,\sigma)$: Find z-score. Use $N(0,1)$ table to find the area = probability

discrete

- **PMF:** $f(x)$ =

  Find prob. by asking which RV's satisfy the range and sum their probabilities

- **CDF:** $F(x)$ =

  Find prob. by drawing Histogram and seeing what mathematical manipulation
  You should use

  i.e. $P(X \leq a) =$ or $P(X < a) =$

  or $P(X \geq a) =$ or $P(X > a) =$

  or $P(a < X < b) =$ or $P(a \leq X \leq b) =$

  or $P(a \leq X < b) =$ or $P(a < X \leq b) =$

- **Binomial:** (5 questions). Use JMP to find the PMF or CDF

86   Introduction to Statistics

Chapter 3 Review Homework:

1. Consider the following probability density function for the continuous random variable, X.

   a. Find h
   b. Find $P(X > 1)$
   c. Find $P(X = ½)$

2. Suppose the random variable X has a normal distribution with a mean of ninety and a standard deviation of six.

   a. Find the Quartiles
   b. Find P_{30}

3. Given the CDF, $F(x) = \begin{cases} 0 & X < 1 \\ \dfrac{x^3 - 1}{26} & 1 \leq X \leq 3 \\ 1 & X > 3 \end{cases}$

   a. Find $P(X < 2)$
   b. Find $P(1.5 < X < 2.5)$
   c. Find $P(¼ < X < ¾)$

4. The recovery time taken for shoulder surgery for a professional pitcher is normally distributed with mean 7 months and standard deviation 1.5 months.

   a. Find the probability that a randomly selected professional pitcher recovers in less than 6 months.
   b. What is the probability that a randomly selected professional pitcher recovers between 5 and 12 months?
   c. Find the length of time in months such that 10% of professional pitchers take less than that time to recover.

5. Suppose the amount of weight in grams that a newborn gains in its first two weeks of life is uniformly distributed over the range of 30 grams to 420 grams. Find the probability that a newborn gains at least 100 grams in its first two weeks of life.

SAMPLING DISTRIBUTIONS

➡ Section 1: Sampling Distributions:

- Recall, Probability Distributions describe _____ possible values that can be obtained by the RV.
- In Chapter 4, our RV is a statistic.
- Recall, from Chapter 1:
  - *Parameter*: A numerical value based on the _____.
  - *Statistic*: A numerical value based on the _____.
- We use the _____ to draw conclusions about the value of the _____.
- Statistics are random variables, i.e., they will vary from sample to sample. Thus, they can be described by their probability distributions.

Definition:

- *Sampling Distribution*: The sampling distribution is the distribution of all possible values that can be assumed by some statistic, computed from samples of the same size randomly drawn from the same population. In other words, it is the probability distribution of the sample statistic. The sampling distribution describes ALL possible values that can be assumed by the statistic.

- The sampling distribution describes the _____.

  - Steps in constructing a sampling distribution:
    1. Randomly draw all possible samples of size *n* from a finite population of size *N*. (If population is infinite, then do "repeated sampling.")
    2. Compute the statistic of interest for each sample.
    3. List in one column the different observed values of the statistic, and in another column the corresponding frequency of occurrence of each distinct value of the statistic. *Or* Construct the histogram – the PDF will be the curve that is attained by smoothing out the histogram.
  - We will be interested in three characteristics about a given sampling distribution:
    1. mean ⟶ What is the "typical value" of the statistic?

2. variance ⟶ Are the values of the statistic similar to each other or are there a lot of different values of the statistic?

3. shape (functional form) ⟶ Is the distribution of the statistic symmetric or skewed? Is it a "well-defined" distribution?

➡ Section 2: The Sampling Distribution of the Sample Mean:

Many times we are interested in probabilities about a group instead of an individual random variable. Thus we need to know the distribution of the statistic. The statistic of interest right now is the sample mean.

- Recall:

  - _____ = population mean.

  - _____ = sample mean.

  - The difference between X and $\overline{X}$:

    _____ = an _____ or a _____ object.

    _____ = the _____ of _____ randomly selected individuals or objects.

- To draw conclusions about the population mean, we need to be able to calculate probabilities about the sample mean. Thus, we need to know the distribution of the sample mean. In other words, we need to construct the sampling distribution of $\bar{x}$.

- A Simulation for Constructing the Sampling Distribution of $\bar{x}$

  To illustrate the distribution of the sample mean, suppose we have the following population. Our random variable X has a Beta(2, 2) distribution. The graph of the PDF is below. By looking at the graph, it is clear that although our population is symmetric, it is not normal.

Following the steps for constructing sampling distributions, from this population, let's take 1000 samples of size 10 and for each sample, calculate $\bar{x}$. (This will give us 1000 $\bar{x}$ values.) This gives us the following graph:

If we compare the graph of the population and the graph of the 1000 $\bar{x}$ values for $n = 10$, what do we notice?

Answer:

Now let's suppose we have the following population. Our random variable X has an Exp(½) distribution. The graph of the PDF is below. By looking at the graph, it is clear that our

population is most definitely not symmetric, in fact it is _____ and thus definitely not normal!

Sampling Distributions **91**

Again, following the steps for constructing sampling distributions, from this population, let's take 1000 samples of size 10 and for each sample, calculate $\bar{x}$. (This will give us 1000 $\bar{x}$ values.) This gives us the following graph:

Sampling Distribution of $\bar{X}$, n=10

If we compare the graph of the population and the graph of the 1000 $\bar{x}$ values for $n = 10$, what do we notice?

Answer:

Now let's take 1000 samples of size 30 and for each sample, calculate $\bar{x}$. (This will give us 1000 $\bar{x}$ values.) This gives us the following graph:

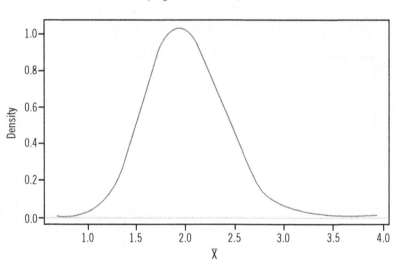

Sampling Distribution of $\bar{X}$, n=30

If we compare the graph of the population and the graph of the 1000 $\bar{x}$ values for $n = 30$, what do we notice?

Answer:

Now let's take 1000 samples of size 150 and for each sample, calculate $\bar{x}$. (This will give us 1000 $\bar{x}$ values.) This gives us the following graph:

If we compare the graph of the population and the graph of the 1000 $\bar{x}$ values for $n = 150$, what do we notice?

Answer:

As you can see from the graphs, sometimes n can be only 10 for the distribution of the sample mean to be "symmetric and bell shaped." Other times, n needs to be 150 before the symmetric bell-shaped curve is obtained.

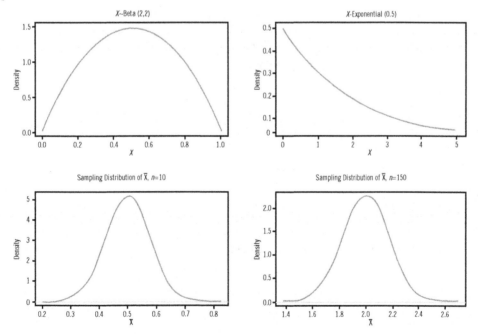

Sampling Distributions 93

- Central Limit Theorem:

  In our sampling distribution simulations, we noted that as n increases, the distribution of $\bar{x}$ becomes more bell-shaped. This happens whether the population distribution is symmetric or skewed. Thus, for a sufficiently large n, the distribution of $\bar{x}$ appears to be normally distributed. As it turns out, it *is* approximately normally distributed, which is given to us via the Central Limit Theorem.

- In Summation: (Properties of the sampling distribution of $\bar{x}$)

  1. The mean of the sampling distribution of $\bar{x}$: $\mu_{\bar{x}} = \mu$.

  2. The variance of the sampling distribution of $\bar{x}$: $\sigma^2_{\bar{x}} = \sigma^2/n$

  3. What about the shape of the distribution of $\bar{x}$?

     Theorem: If a random sample of n observations is selected from a normal population, the sampling distribution of $\bar{x}$ will also have a normal distribution.

     Central Limit Theorem (CLT): Suppose a random sample of size n is selected from *any* population. When n is sufficiently large, the sampling distribution of $\bar{x}$ has an approximate normal distribution. As n gets larger, this approximation becomes better.

- What does "sufficiently large" mean? Most elementary textbooks suggest the minimum is $n \geq 30$. *This is JUST a rule of thumb*. The required size of n for $\bar{x}$ to be approximately normal depends on the population from which one took samples. If the population is highly skewed, then n may need to be quite a bit more than thirty, possibly up to $n = 150$ (as we saw). If the population is symmetric and bell-shaped, then n may be able to be much smaller than thirty, possibly as small as $n = 10$ (again as we saw). Thus, in practice, one should always look at a histogram of the data to verify normality, i.e., to verify that a sufficiently large enough sample was obtained.

- Examples

  1. On Texas Interstate Highways the speed of cars is a normally distributed random variable with mean of seventy mph and standard deviation of five mph.

     a. What does the random variable, X, represent?

     b. Do we know the distribution of X?

     c. What does the random variable, $\overline{X}$, represent?

94   Introduction to Statistics

d. Determine the sampling distribution of the sample mean for samples of size twenty.

e. What is the probability that the mean speed in a random sample of twenty cars is within two mph of the population mean speed of seventy mph.

2. Assume that cans of Dr. Pepper are filled so that the actual amounts have a mean of 12.00 oz. and a standard deviation of 1.2 oz.

a. What does the random variable, X, represent?

b. Do we know the distribution of X?

c. What does the random variable, $\overline{X}$, represent?

d. Determine the sampling distribution of the sample mean for samples of size thirty-six.

e. Find the probability that a sample of thirty-six cans will have a mean amount of at least 12.62 oz.

Sampling Distributions **95**

Checking your Understanding: (Answers are found in the Appendix)

1. Suppose it is known that adolescents play an online game for an average of 81 minutes per day with standard deviation of 22 minutes. If a random sample of 40 adolescents is polled, find the probability that the mean minutes spent playing an online game per day is less than 92 minutes.

2. Suppose the life spans of treadmills are known to be normally distributed with a mean of 8.25 years and a standard deviation of 3.25 years.

   a. What is the probability that the mean life span in a group of 10 randomly selected treadmills is between 7 and 11 years?

   b. What is the probability that a randomly selected treadmill has a life span of less than 13 years?

Section 2: Homework

1. Suppose the random variable X is drawn from a population with mean thirty and standard deviation five.

   a. For samples of size sixty, what is the distribution of the sample mean?

   b. Is the answer to part a dependent upon the sample size being at least a specific size?

2. Suppose the random variable X is drawn from a normal population with mean fifty and standard deviation two.

   a. For samples of size sixty, what is the sampling distribution of the sample mean?

   b. Is the answer to part a dependent upon the sample size being at least a specific size?

   c. Is the answer to part a dependent upon knowing the distribution of X?

3. The average arm span for females is 63.75 inches with standard deviation 3.4 inches. Certain athletes, such as swimmers, rowers, or boxers benefit from having a large arm span. Suppose forty female swimmers are randomly selected. What is the probability that their average arm span is greater than 65.8 inches?

4. According to the American Music Industry, the length of a song is normally distributed with a mean of 240 seconds and standard deviation thirty seconds. Suppose a random sample of fifteen songs is selected from the morning program. What is the probability that the mean length of the song is between 225 and 260 seconds long?

5. The recovery time taken for shoulder surgery for a college fast-pitch pitcher is normally distributed with mean 8 months and a standard deviation of 1.2 months. If you randomly select 16 college fast-pitch pitchers, what is the probability that their mean recovery time is less than 8.5 months?

6. Suppose the average time employees spend in an online work meeting per day is 2.5 hours with standard deviation 1.05 hours. Further suppose Aaron's Inc. has forty employees. Find the probability that the average time they spend in an online work meeting per day is between 2 and 3 hours.

Section 3: The Sampling Distribution of the Sample Proportion:

Sometimes we are interested in knowing the percentage of subjects that have a certain characteristic of interest. Or we might want to determine who the preferred presidential candidate will be in the upcoming election. In these cases our data is categorical, i.e., counting the number of subjects with the specified characteristic or counting the number of votes received by a candidate. For both examples, the parameter(s) of interest is the population proportion.

- p = population proportion.
- $\hat{p}$ = sample proportion.
- $\hat{p} = \dfrac{x}{n}$, where x = the # of those with the characteristic and n = the total sample size.
- The difference between X and p and $\hat{p}$

_____ = the _____ (or objects) with the characteristic of interest.

_____ = the _____ of the _____ with the characteristic of interest.

_____ = the _____ of the _____ with the characteristic of interest.

- The Sampling Distribution of $\hat{p}$. To find the sampling distribution, we need to know the following:

  1. The mean of the sampling distribution of $\hat{p}$: $\mu_{\hat{p}} = p$

  2. The variance of the sampling distribution of $\hat{p}$: $\sigma_{\hat{p}}^2 = \dfrac{pq}{n}$

  3. What about the shape of the distribution of $\hat{p}$?

  Recall that when we are interested in the population proportion, our data generally consists of count data. In other words, we let our random variable X count the number of subjects with the desired characteristic, etc. In order to determine the sampling distribution of $\hat{p}$, we define our random variable as such:

  $$\text{Define } x_i = \begin{cases} 1 \text{ if the subject has the characteristic of interest} \\ 0 \text{ if the subject does not have the characteristic of interest.} \end{cases}$$

Sampling Distributions **97**

Thus we can define $\hat{p}$ to be

$$\hat{p} = \frac{\sum x_i}{n}$$

- By looking at the formula for $\hat{p}$, what can we say about the functional form of $\hat{p}$?

  Answer:

- What are the assumptions?

  Answer:

- Examples

  1. Mayor Alice Cook is running for reelection and believes that she can win the election if she can poll a majority of the votes in the Northern precinct. She also believes 55% of the city's voters favor her. If 150 voters show up in the Northern precinct, what is the probability Mayor Alice Cook receives a majority of the Northern precinct votes?

2. The Texas High School Coaches Association claims that 12% of high school athletes have suffered a career ending injury while playing a high school sport. Out of 100 randomly chosen high school athletes, what is the probability that less than 7% will have a career ending injury while playing a high school sport?

The Relationship between $\hat{p}$ and the Binomial Distribution:

In order to see how the size of n affects the shape of the Binomial distribution, let's consider the Binomial distribution with $p = 0.1$.

Below is the histogram for the RV X where $X \sim \text{Bin}(10, 0.1)$:

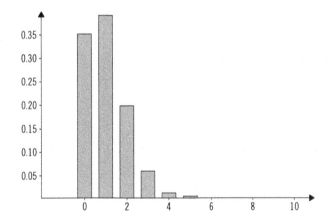

The Shape of the distribution is: _____.

Now let's look at the histogram for the RV X where $X \sim \text{Bin}(150, 0.1)$:

What do you notice?

This illustrates for us that the Binomial distribution for a sufficiently large n is approximately Normal. Thus, when X is Binomial, if n is large enough, instead of solving the Binomial problem in terms of X, we can convert X to $\hat{p}$ and solve the problem as a Normal distribution!

3. The American Restaurant Association finds that 8% of persons who make reservations do not show up. Suppose Andy's Five-Star Cuisine Restaurant has 160 tables and reservations are made for all tables for Friday night. Further suppose that ten couples, who don't have reservations, show up to eat. What is the probability that all the couples will be able to eat at their own table?

- Example 3 illustrates for us that a binomial random variable for large n, is approximately normal. Below is the histogram of the random variable from example 3.

Notice that the histogram for the binomial distribution above fits quite nicely under the normal curve. Thus we can convert the x to a $\hat{p}$ and approximate the problem using the normal distribution.

Checking your Understanding: (Answers are found in the Appendix)

1. Suppose that 35% of all high school students take shop class. If you sample 10 high school students, what is the probability that more than 30% of the students you sample take shop class?

2. Suppose that 30% of all retired people take an online class. If you sample 100 retired people, what is the probability that more than 35% of the them are taking an online class?

Section 3: Homework

1. Suppose the proportion of American adults who work at home is 40%. For a random sample of 100 American adults, what is the distribution of the sample proportion?

2. According to the Federal Insurance Agency statistics, the proportion of teenage drivers between the ages of 16 to 19 who have received a speeding ticket is 54%. From a random sample of one hundred teenage drivers, what is the probability that a majority of them have received a speeding ticket?

3. Suppose the proportion of elementary students in Texas that are enrolled in fully remote learning programs is 20%. What is the probability that at most 10% of 100 randomly selected elementary students from Texas is enrolled in a fully remote learning program?

4. The proportion of college students that participate in a work-study program is 60%. From a random sample of fifty college students, what is the probability that at least forty of them participate in a work-study program?

5. A manufacturer of cell phones purchases batteries from a vendor. It is not uncommon to receive defective batteries from this vendor, thus whenever a shipment is received, a sample of batteries is inspected. For this month's shipment, a random sample of 280 batteries is selected and each battery is inspected. If the proportion of defective batteries sampled is more than 1.5%, the entire shipment will be returned to the vendor. What is the probability that this month's shipment will be returned if the true proportion of defective batteries is 4%?

▶ Section 4: Probability Distributions Overview:

When you see the key word "probability," your first question should be, "Is my probability about a random variable X or about a statistic?" If it is about X, then we are interested in a _____ or

_____. If it is about a statistic, then we are interested in a _____

and thus given a _____.

Sampling Distributions 101

- *X*?

102 Introduction to Statistics

• *A Statistic?*

Sampling Distributions **103**

Chapter 4 Review Homework:

1. Suppose the random variable X is drawn from a population with mean fifty and standard deviation three. Is the sampling distribution of the sample mean for a sample of size twenty known or unknown? If known, what is it?

2. Suppose $X \sim Bin(26, 0.5)$. Further suppose a researcher is interested in calculating the probability that a majority of the trials are a success. Can the researcher calculate the probability in terms of the sample proportion or must he solve it in terms of the random variable X?

3. Fifty-one percent of adults in the United States whose New Year's resolution was to exercise more achieved their resolution. Suppose a random sample of sixty-five American adults whose New Year's resolution was to exercise more is chosen. What is the probability that a majority of them achieved their resolution?

4. Fifteen percent of adults in the United States do not make New Year's resolutions. Suppose a random sample of sixteen American adults is chosen. What is the probability that a majority of them do not make New Year's resolutions?

5. The time in seconds that it takes a Baylor student to jog from the Student Union Building to the Student Life Center is distributed as a Normal random variable with mean ninety-five seconds and standard deviation of fifteen seconds.

    a. What is the probability that a student runs to the Student Life Center in at most two minutes?

    b. What is the time, in seconds, at which the fastest 20% of students will have completed the run?

    c. Suppose fifteen students decide to race from the Student Union Building to the Student Life Center. What is the probability that the average time it takes the students to get to the Student Life Center exceeds 100 seconds?

    d. Suppose fifteen students decide to race from the Student Union Building to the Student Life Center. What is the probability that the average time it takes the students to get to the Student Life Center exceeds 150 seconds?

6. Suppose that 21% of men in the United States consider fishing their favorite leisure-time activity. A random sample of ten men is selected. What is the probability that at least 20% of them will say that fishing is their favorite leisure-time activity?

7. The incubation period for the eggs of a house wren is normally distributed with a mean of 336 hours and a standard deviation of 3.5 hours.

    a. Find the probability that an egg has an incubation period between 328 and 338 hours.

    b. Suppose a researcher has fifteen house wren eggs in an incubator. Find the probability that the average incubation time for the fifteen eggs is greater than 337 hours.

    c. Suppose it is known that 33% of House Wren eggs never hatch. Suppose an ornithologist has forty house wren eggs. Find the probability that less than 25% never hatch.

8. Suppose the average GPA for students nationwide that are enrolled in only online courses is 3.10 with standard deviation 0.30. What is the probability that a random sample of forty nine students that are enrolled in only online courses has an average GPA greater than 3.20?

Sampling Distributions **105**

ESTIMATION AND CONFIDENCE INTERVALS FOR ONE PARAMETER

Statistical Inference – the process of making judgments about a population based on properties of a sample randomly drawn from the population.

Estimation – involves approximating the value of an unknown parameter.

Hypothesis Testing – involves choosing between two opposing statements concerning a parameter

➡ Section 1: Point Estimate:

Point estimate – when we use a statistic to estimate the value of a parameter.

- Suppose I ask one hundred randomly selected Baylor students how much money they spent eating out last month. The sample mean for the one hundred students is $182.25.

   If I make the following statement: "*A random sample of 100 Baylor students spent $182.25 eating out last month,*" am I describing the population or the sample?

   Then I have simply made a _____ statement.

   If, however, I make the statement; "*Baylor students spent an average of $182.25 eating out last month,*" am I now describing the population or am I still describing the sample?

   Then I have made an _____ statement. I am using information from the

   _____ to describe the _____ .

   In other words, I am using the _____ as an estimate of the

   _____ .

- We define the sample mean, $\bar{x}$, to be a *point estimate* of the population mean, μ.
- We define the sample proportion, $\hat{p}$, to be a *point estimate* of the population proportion, p.

In the above example, I am interested in the *typical* amount Baylor students spend on eating out per month. Recall, this is what we refer to as a Measure of Central Tendency. What are the three Measures of Central Tendency?

Answer:

Estimation and Confidence Intervals for One Parameter **107**

Recall that sometimes it is best to use one statistic versus another when describing the typical data value. How do we know which statistic to use? We know that if our data is skewed or contains outliers, we should use the _____. However, the mean is the most commonly one used because of the following:

- When choosing which statistic to use to estimate the parameter, we want to choose the statistic that has the following two properties:
   1. The statistic is *unbiased*.
   2. The statistic has *minimum variance*.

      Let's first define what we mean by unbiased: Consider the following distributions:

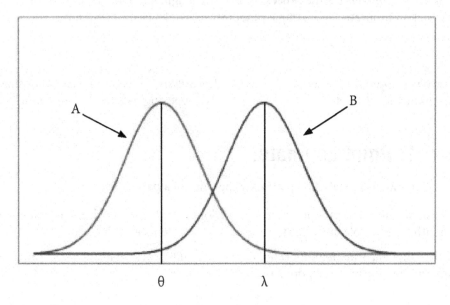

- Assume θ is the population parameter that we wish to estimate. Also, assume both $\hat{\theta}$ and $\hat{\lambda}$ are estimates of θ.
- Let A be the distribution of $\hat{\theta}$. Thus, the curve A describes:
   Answer:

- Let B be the distribution of $\hat{\lambda}$. Thus, the curve B describes:
   Answer:

108   Introduction to Statistics

- By looking at the curve A, we know the E[$\hat{\theta}$] = _____ . This implies:

    Answer:

- By looking at the curve B, we know the E[$\hat{\lambda}$] = _____ . This implies:

    Answer:

- The statistic $\hat{\theta}$ is *unbiased* if the mean of the sampling distribution of $\hat{\theta}$ is equal to the parameter, θ. i.e., $\hat{\theta}$ is an unbiased estimator of θ if E[$\hat{\theta}$] = θ.
- We say $\hat{\lambda}$ is a *biased* estimator of θ, but an unbiased estimator of _____ .

• Let's now define what we mean by minimum variance. Consider the following distributions:

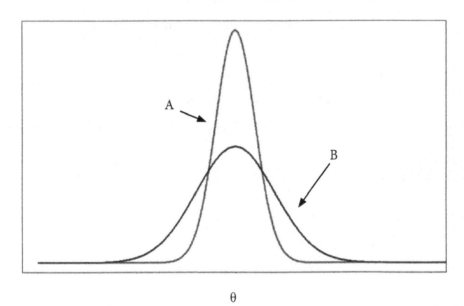

- Let A be the distribution of $\hat{\theta}$ and B be the distribution of $\hat{\delta}$.
- E[$\hat{\theta}$] and E[$\hat{\delta}$] = _____ , thus both $\hat{\theta}$ and $\hat{\delta}$ are _____ estimators of θ.
- Label on the graph which range of values of $\hat{\theta}$ are most likely to occur.
- Label on the graph which range of values of $\hat{\delta}$ are most likely to occur.
- The statistic $\hat{\theta}$ said to have *minimum variance* if the sampling distribution of the statistic has the smallest variance of all the unbiased estimators.

Estimation and Confidence Intervals for One Parameter

Why is this good thing?

Answer:

A Simple Explanation of Unbiased and Minimum Variance:

Suppose Ana, Ava, and Ali have lost their sister, Adi, on the playground. Since she is young, they don't want their mom to know they have lost her, so they want to find her as quick as possible!

Adi is the unknown parameter!

Ana searches first, but does not find her. She gets "close" to finding her but ends up at the slide.

Ana chose a biased path.

Ali searches next for Adi. She goes by the teter-totter and then loops around the big tree, before eventually finding Adi at the swingsets. However, she did not find her before her Mom knew she was missing.

Ali chose an unbiased path, but not the path with minimum variance.

Ava searches for Adi by going straight to the swingsets. She finds Adi before her Mom knows Adi is missing.

Ava chose the unbiased path with minimum variance.

Ava was the "best estimator" for finding Adi, the "unknown parameter"!

110   Introduction to Statistics

Now back to our Baylor eating out example:

Do you think it is very likely that the true value of the population mean is $182.25? In other words, do you think it is true that the average amount that ALL Baylor students spent on eating out last month is $182.25? Most likely not. It *could* be the case that the one hundred students selected just happened to spend more on eating out than the other Baylor students. In fact, the average for all Baylor students (the *population mean*) could be very different from the *sample mean* of $182.25.

Unless we perform a census, we can never know with absolute certainty even approximately what the true population mean is. For instance, what if one student not polled happened to spend $1,000 on eating out last month? We know this outlier will drastically affect the sample mean!

Can we improve our estimate of the population mean? **Yes!**

Instead of just using a point estimate, can we do better? **Yes!**

Checking your Understanding: (Answers are found in the Appendix)

Researchers are interested in the proportion of students that are required to take an advanced level course in statistics prior to being admitted to nursing school. To determine, a random sample of 100 pre-nursing students is obtained and of those sampled 56 are required to take an advanced level course in statistics in order to gain admittance into nursing school.

a. What is the parameter of interest?

b. What is the point estimate you would calculate to estimate the parameter? What is its value?

c. Is the point estimate an unbiased estimator of the parameter? Why or why not?

Section 1: Homework:

1. When choosing a point estimate, we want to choose the point estimate that has what two properties?

2. Is $\bar{x}$ an unbiased estimator of μ? Why or why not.

3. Is $\hat{p}$ an unbiased estimator of p? Why or why not.

4. What is the point estimate for p?

5. What is the point estimate for μ?

6. The ASA softball association is interested in knowing the average pitching speed of a 12U fastpitch softball pitcher. A random sample of forty pitches is recorded. The average speed of the pitch is found to be forty-one mph with standard deviation two mph.

   a. What is the parameter of interest?

   b. What is the point estimate you would calculate to estimate the parameter? What is its value?

7. Across the nation, many college students have been complaining to the dean of students about the temperature in the classrooms being at an "uncomfortable level", either too cold or too hot. Baylor's dean of students is interested in determining the true proportion of Baylor students that feel the classroom temperature is at an uncomfortable level. Out of a random sample of 200 Baylor students, ninety-eight say that the classrooms are either too hot or too cold.

   a. What is the parameter of interest?

   b. What is the point estimate you would calculate to estimate the parameter? What is its value?

Estimation and Confidence Intervals for One Parameter **111**

Section 2: Confidence Intervals:

A *confidence interval* is an interval estimate of the parameter in which we include how "confident" we are that the interval contains the parameter we are estimating.

> Instead of using one point to estimate the parameter, we use a range of values as the estimate, making sure we include how "confident" we are that the interval contains the true value of the parameter.

- What do we mean when we talk about how "confident" we are that the range contains the parameter?

> A 95% confidence interval is an interval generated by a process that's right 95% of the time. Similarly, a 90% confidence interval is an interval generated by a process that's right 90% of the time and a 99% confidence interval is an interval generated by a process that's right 99% of the time.
>
> Another way to say this is that if we were to replicate our study many times, each time reporting a 95% confidence interval, then 95% of the intervals would contain the true value of the parameter.
>
> As an illustration, suppose we are drawing from a population in which we know the population mean is equal to fifty. We are going to construct ten different 90% confidence intervals from this population. The picture below illustrates what we mean by 90% confidence.

Two things to notice:

1. How many of the confidence intervals contain the true value of µ? How many do not?

2. Are the endpoints for each confidence interval the same or different?

- Comments:

  1. If we construct a 90% confidence interval for µ to be, $5 < \mu < 15$, does that imply that 90% of the time, the value of µ is between *5* and *15*?

     Answer:

  2. In practice, we perform our study only once. i.e., we construct one confidence interval. We have no way of knowing whether our particular interval is correct, but we behave as though it is.

  3. *The amount of confidence affects the length of the interval. Keeping all else the same, if we increase the confidence of the interval, we increase the length of the interval.*

  4. In theory, we can construct intervals of any level of confidence from 0 to 100%. The longer the interval, the more likely it is to contain the quantity being estimated, but the length could be too long to be useful in estimation.

5. Ninety-five percent has been found to be a convenient level for conducting scientific research, so it is used almost universally. Intervals of lesser confidence would lead to too many misstatements. Greater confidence would require more data to generate intervals of usable lengths.

- How do we construct a confidence interval (CI)?

When constructing a CI, we begin with a point estimate and "stretch" that point estimate into an interval estimate (or range of values) in a way that allows us to quantify our "uncertainty" about that interval containing the true value of the parameter.

The amount in which we stretch the point estimate is called the Margin of Error and is denoted by E.

This gives us the following general formula for finding a CI.

$$\text{Pt. Est.} \pm E$$

$$(\underline{\text{Pt. Est.} - E}, \underline{\text{Pt. Est.} + E})$$

Thus, we see that the length of the confidence interval, $L =$ _____. Which

implies the Margin of Error, $E =$ _____.

- When estimating a parameter, what "type" of point estimate do we want to use?
  Answer:

- What statistical tool do we use to quantify our uncertainty?
  Answer:

- What describes for us all possible values of a point estimate/statistic?
  Answer:

Thus, when calculating a CI, we are finding the points a and b on the distribution curve of the point estimate such that the probability of being between a and b, centered about the mean of the distribution, is equal to our level of confidence.

Estimation and Confidence Intervals for One Parameter

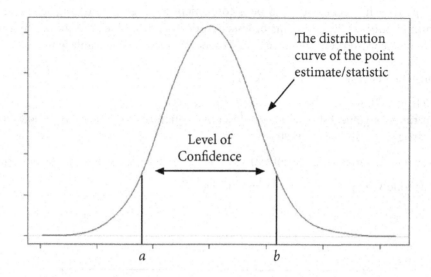

Two points to understand:

1. The curve we are using is the _____ curve of the point estimate and describes for us _____ possible values that can be obtained by the _____ .

2. The mean of the distribution is equal to the _____ we are estimating.

Checking your Understanding: (Answers are found in the Appendix)

1. Suppose $E = 10$. Find the length of the confidence interval.
2. Suppose the length of the confidence interval is 9 and the point estimate is 12. Find the confidence interval.

Section 2: Homework:

1. What is the goal of a confidence interval?
2. Suppose the margin of error for estimating a parameter is ten. What is the length of the resulting confidence interval?
3. Suppose the length of a confidence interval is eight. What is the Margin of Error?
4. Suppose the length of a confidence interval is ten. If the point estimate is forty-five, what is the confidence interval?
5. Suppose the margin of error of a confidence interval is six. If the point estimate is eighty, what is the confidence interval?
6. Keeping all else the same, is a 95% confidence interval narrower or wider than a 90% confidence interval?
7. Keeping all else the same, is a 95% confidence interval for $n = 100$ narrower or wider than a 95% confidence interval for $n = 200$?
8. Suppose the confidence interval equals (12, 18). What is the value of the point estimate?

▶ Section 3: A (1 − α)% Confidence Interval for p:

Review:

- What is the point estimate of p?

- What is the sampling distribution (including assumptions)?

- Thus, we know the _____ curve centered about _____ describes for us _____ possible values of _____ .

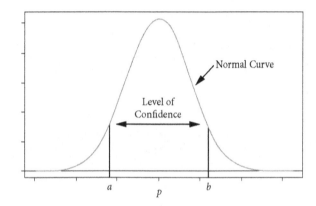

- If we want to find the points a and b on this curve, what must we find first?

- If I want to find a 90% CI, the _____ for the points a and b would be?

- This implies that the points a and b are _____ standard deviations above and below the mean of the distribution. What do I do to determine how many units the points a and b are above and below the mean of the distribution?

Estimation and Confidence Intervals for One Parameter   115

- Hence, the formula for finding a 90% CI for p is:

$$\underline{\hspace{3cm}} \pm \underline{\hspace{3cm}} \cdot \underline{\hspace{3cm}}$$

      Pt. Est. for p The z-score   The std. dev. of
                             (from the dist.   the pt. est.
                             of the pt. est.)

- What is wrong with this formula? What is the goal for finding a CI?

- Two ways to handle this problem:

  1. A conservative approach:

  2. Use previous information:

- The general form of the CI for p:

$$Pt.\ Est. \pm E$$

$$Pt.\ Est. \pm RF \cdot SE$$

The one point you   Dist. Of   The std. dev
use to estimate   the *Pt. Est.*   Of the *Pt. Est.*
the parameter

Assuming:

- Example 1: A poll was taken of 1200 U.S. nurses. The nurses sampled were asked whether they have ever administered an incorrect amount of medicine to a patient and 240 responded "yes." Use these data to obtain a 95% confidence interval for the proportion of all US nurses who have made an error in administering medicine to a patient.

  - Interpretation:

  - Margin of Error, E:

Estimation and Confidence Intervals for One Parameter **117**

- Interpretation of E:

- Can we be 95% confident that more than 20% of US nurses have made an error in administering medicine?

- Can we be 95% confident that less than 25% of US nurses have made an error in administering medicine?

- For a CI to support a claim, _____ value within the CI has to support the claim!

- Let's make sure we really understand what a CI implies and/or doesn't imply. Answer the following True/False Questions:

  1. The above implies that 95% of the time the proportion of US nurses who have made an error in administering medicine is between 17.74% and 22.26%.

     A   True   B   False

  2. The above implies that $P(.1774 < p < .2226) = 0.95$

     A   True   B   False

3. Saying we are 95% confident implies that the method used to construct the interval will result in an interval, 95% of the time, which contains the true proportion of US nurses who have made an error in administering medicine.

 A   True   B   False

4. We have no idea if the true proportion of US nurses who have made an error in administering medicine is between 17.74% and 22.26%, but we are 95% confident that it does, therefore we use these as valid estimates of the true value of p.

 A   True   B   False

Checking your Understanding: (Answers are found in the Appendix)

Suppose the administrator in charge of student tickets is interested in estimating the proportion of students who left before half time at the last football game. This administrator randomly selects 60 students who attended the last game and asks if they left before half time. Of the 60 students, 32 said yes, they did.

a. Find a 95% confidence interval for the true proportion of students who left before half time at the last football game.

b. Can the administrator claim with 95% confidence that a majority of the students left before half time of the last game? Explain your answer.

Section 3: Homework:

1. Suppose your dorm refrigerator is no longer working and you want to buy a new one. Your roommate thinks it would be cheaper to have yours repaired. To settle the argument, a random sample of ten dorm refrigerators is taken and the resulting 95% confidence interval for the mean repair cost is (59.48, 90.52). Based on the confidence interval, can you be 95% confident that it will cost more than $62 to repair the dorm refrigerator?

2. A 95% confidence interval for the proportion of Americans under the age of thirty who skip breakfast is found to be (0.2025, 0.2795). For each choice below, answer True or False:

 a.   T   or   F   There is a 95% probability that the true proportion of Americans under the age of thirty who skip breakfast is between 20.25% and 27.95%.

 b.   T   or   F   If this study were to be repeated many times with samples of the same size, then 95% of the time the true proportion will be between 0.2025 and 0.2795, on average.

 c.   T   or   F   We can be 95% confident that, on average, the proportion of Americans under the age of thirty sampled who skip breakfast is between 0.2025 and 0.2795.

 d.   T   or   F   We can be 95% confident that, on average, the proportion of Americans under the age of thirty who skip breakfast is between 0.2025 and 0.2795.

 e.   T   or   F   If we were to construct one hundred confidence intervals, then 95% of them would contain the true proportion of Americans under the age of thirty who skip breakfast, five percent would not.

Estimation and Confidence Intervals for One Parameter **119**

3. Suppose a random sample of size 60 is obtained and the sample proportion is found to be 0.25.

    a. Find a 95% confidence interval for the true proportion.

    b. Would a 99% confidence interval be wider or narrower than the confidence interval found in part a?

    c. Suppose a sample size of 50 is obtained instead of size 60, would the resulting 95% confidence interval be narrower or wider than the confidence interval found in part a?

4. One of the hot topics in the last presidential election was with respect to our citizen's health and whether or not the government has a right to limit our choices in what we eat or drink. A random sample of 1046 Americans was obtained. The question asked was "Do you support our government banning specific types or sizes of food or drinks?" Of those surveyed, 895 responded, "no."

    a. Estimate with 95% confidence the proportion of Americans who do not support the government banning specific types or sizes of food or drinks.

    b. Using the confidence interval found in part a, can you conclude that a majority of Americans do not support the government banning specific types or sizes of food or drinks. Why or why not.

5. The governor of New York has banned selling or purchasing "large" soft drinks, where a "large" is defined as being more than 16 oz. Sonic, famous for its Route 44 drinks is interested in estimating the proportion of New Yorkers that favor this ban. From a random survey of 800 New Yorkers, 375 answered no to the question, "Do you favor the governor's ban on 'large' soft drinks?"

    a. Find a 97.5% confidence interval for the proportion of New Yorkers who favor the ban on "large" soft drinks.

    b. Interpret the confidence interval.

    c. Can you conclude with 97.5% confidence that a majority of New Yorkers favor the ban on "large" soft drinks? Why or why not.

    d. Without doing any other calculations, can you be 99% confident that a majority of New Yorkers favor the ban on "large" soft drinks?

➡ Section 4: Determining n:

In theory a researcher can construct a CI at any level of confidence. But, we also know that holding all else the same, if we increase the level of confidence, the length of the CI will increase. Our goal is to have the maximum level of confidence with the minimal (usable) length.

In many studies, the researcher wants to be able to estimate the parameter with a specific level of confidence and to be within a specified value. In order to do this, the researcher needs to be able to determine the sample size needed to insure this certain degree of accuracy in estimation.

There are several factors to consider in doing this:

1. Level of *Confidence* – refers to how "confident" we are that the interval actually contains the true value of the parameter.

2. *Precision* – refers to the width of the confidence interval.

3. *Variability* – refers to the variability of the population. The more variability in the population the larger the sample size needs to be.

120   Introduction to Statistics

4. *Complexity of Analysis* – the more variables that one is analyzing simultaneously, the larger the sample size needs to be.

5. *Resources* – larger sample sizes require more time and money.

Determining the sample size needed when estimating a proportion:

- Recall the formula for finding a CI for p is

$$\hat{p} \pm Z_{\alpha/2} \cdot \sqrt{\frac{\hat{p} \cdot \hat{q}}{n}}$$

where the Margin of Error equals

$$E = Z_{\alpha/2} \cdot \sqrt{\frac{\hat{p} \cdot \hat{q}}{n}}$$

and represents the precision, which is the amount we want the estimate to be "within" the true value of the parameter. Thus to determine the sample size needed to estimate the population proportion with a given level of confidence and precision, we simply solve the above formula for n.

Formula for finding n when estimating p:

where $n =$ _____ .

$E =$ _____ $=$ _____ .

and $Z =$ _____ .

- What about $\hat{p}$ and $\hat{q}$? Do we know their value?
  Answer:

- What do we do if not?
  Answer:

Estimation and Confidence Intervals for One Parameter **121**

- Example 1: The nursing manager was concerned to see the proportions found in the confidence interval above. She decides to calculate her own confidence interval. What sample size should she take so that she can estimate the true proportion of nurses who have made an error in administering medicine to be within 2% with 99% confidence?

    a. Suppose the point estimate of 20% from the previous example is valid.

    b. Assume there is no information known about the true proportion.

Checking your Understanding: (Answers are found in the Appendix)

Suppose it is believed that 36% of homes in the United States subscribe to the some type of home viewing movie service. We would like to estimate the true proportion of homes in the US that subscribe to some type of home viewing movie service. What sample size should we use to ensure we estimate the truth within 3% with 95% confidence?

Section 4: Homework:

1. What sample size should a researcher take if he wants to estimate a population proportion to be within 2% of the true value with 90% confidence?

2. What sample size should a researcher take if he wants to estimate a population proportion to be within 3% of the true value with 95% confidence? Previous research suggests the proportion is equal to 25%.

3. To estimate the proportion of teen traffic deaths in Texas last year that was caused by speeding, determine the necessary sample size for the estimate to be accurate to within .05 with confidence .99.

    a. Suppose from a previous study, we expect the proportion to be about .30.

    b. Suppose we do not have any prior knowledge about the true proportion.

122 Introduction to Statistics

4. The registrar's office would like to estimate the proportion of students that want to take all on-line courses next semester. What sample size should the registrar's office take so that their estimate is within 2.5% with 95% confidence.

➡ Section 5: A (1 − α)% Confidence Interval for μ:

- The general form for a confidence interval is as follows:

    Point Estimate $\pm$ Margin of Error

    Point Estimate $\pm$ Reliability Factor * Standard Error

- So for μ, we have the following general form of the CI:

    Answer:

    - What is the goal when finding a CI for μ?

        Answer:

    - Do we know the value of μ?

        Answer:

    - How likely is it that we will know the value of σ?

        Answer:

- So, if we don't know σ, what can we use as estimate for it?

- When we replace σ with s, the distribution of the point estimate is no longer normal. It was discovered by Dr. Gosset to have a student's t-distribution (t-distribution) Thus, we have the following formula for finding a CI for μ:

    Answer:

Estimation and Confidence Intervals for One Parameter   **123**

- <u>Formula for a (1 − α)% confidence interval for μ:</u>
  - When σ is known:

  This formula is seldom used in practice.

  - When σ is unknown:

  This formula is the one most often used "in the real world".

  Again, where *t* stand for the student's *t*-distribution with degrees of freedom, $df = n - 1$.

- Properties of the student's *t*-distribution:
  1. symmetric bell-shaped curve
  2. centered at the mean = 0
  3. has more variability than the $N(0,1)$ curve. (fatter tails)
  4. there are infinitely many different *t*-curves. Each one is determined by its degrees of freedom.
  5. as $n \to \infty$, the *t*-curve $\to N(0,1)$.
- In order to use the student's *t*-distribution, our population must be normally distributed. We call this a distributional assumption. Thus, the assumption is that the population from which the data is drawn from must be normally distributed.
- How do we check normality? We have a computer statistical package, such as JMP, create a Normal Quantile Plot, (also called a QQ-Plot).

This is what the QQ plot looks like for data that is normally distributed:

124   Introduction to Statistics

If the data values are forming an overall concave shape, either concave up or concave down, then we cannot assume that the population is normally distributed. In fact, the distribution is skewed.

Over-all concave up implies the population is _____.

Over-all concave down implies the population is _____.

- Example 1: Crime analysts compile data on misdemeanor theft crimes. A theft is considered a misdemeanor if the item stolen is not considered too valuable. A sample of twenty-four of last year's misdemeanor thefts had a mean value loss of $53.63 with standard deviation $25.53.

    a. Obtain a 95% confidence interval for the mean value lost of last year's misdemeanor theft crimes.

    b. Interpret the CI

    c. Find the margin of error:

Estimation and Confidence Intervals for One Parameter 125

d. Interpretation of *E*:

- Suppose the data is given. If we have the data, we can use JMP to find the CI and to check our assumption.

47	34	28	80	96	43	76	55
27	14	106	75	57	29	27	45
88	39	89	54	48	43	65	22

> **JMP Instructions:**
>
> 1. Input data
> 2. Under *analyze*, choose *distribution*
> 3. Click on the column containing the data, then click on *Y*, then hit *okay*.
> 4. Under the red options button next to distribution, choose *stack*
> 5. Under red options button next to column 1 (or whatever you named your column), choose *Normal Quantile Plot*
> 6. Under red options button next to column 1 (or whatever you named your column), choose *confidence interval*
> 7. Choose the correct significance, hit *okay*.

- JMP Output:

126 Introduction to Statistics

Confidence Intervals				
Parameter	Estimate	Lower CI	Upper CI	1-Alpha
Mean	53.625	42.84367	64.40633	0.950
Std Dev	25.53227	19.84403	35.81566	0.950

Checking your Understanding: (Answers are found in the Appendix)

1. Suppose it is known that people talked on their home phone for an average of 21 minutes per day. Researchers want to know if having such easy access to cell phones has changed the amount of time people talk on a phone. Thus, they would like to estimate with 95% confidence the average number of minutes that people talk on a phone (cell or land). Suppose a random sample of 40 people is polled, and the mean number of minutes talked on a land line or cell phone is 24 minutes with standard deviation 5 minutes.

   a. Estimate with 95% confidence the average number of minutes that people talk on a phone (cell or land)

   b. Can you be 95% confident that the average number of minutes people talk on a phone (cell or land) is more than 21 minutes? Explain your answer.

2. Suppose a city architect is interested in the heights of the buildings in his town. He would like to know if the average height of the buildings in his town is different than 105 ft. The architect randomly selects 10 buildings and their heights, in feet, are listed below.

   106   109   107   108   112   113   104   105   115   114

   a. Find a 99% confidence interval for the average height of the buildings.

   b. Can the city architect conclude with 99% confidence that the average height is different than 105 ft.

Section 5: Homework

1. With economical concerns being so great, many people are no longer purchasing new vehicles, but are instead keeping their vehicles for a longer period of time, until they no longer work. The American Automobile Association is interested in estimating the average age of vehicles for Americans. A random sample of thirty-five Americans was obtained. The average age of their vehicles was 6.2 years with standard deviation 1.2. Estimate with 99% confidence the average age of vehicles for Americans.

2. Many people have heard of the "Freshmen 20," referring to the belief that many college freshmen will gain 20 lbs. their freshmen year in school. Administrators were interested in knowing if this is really true. A random sample of thirty freshmen (at the end of their freshmen year) was obtained. The average amount of weight gained by the freshmen is listed below.

   21.8   20.6   21   20.2   19.2   17.8   22.8   20.2   22   19.4

   18.6   18.2   20   19   21   23   19.6   19   24.6   19.6

   17.8   20.6   20.8   17.2   22.4   19.2   23.4   18.2   20.6   21.4

Estimation and Confidence Intervals for One Parameter **127**

a. Find a 97.5% confidence interval for the average amount of weight gained for freshmen.

b. Interpret the CI.

c. Using your CI found in part a, can you conclude with 97.5% confidence that college freshmen gained more than 20 lbs.?

3. Researchers are interested in estimating the average number of credit hours Freshmen take per semester. A random sample of 50 Freshmen is obtained and the average number of hours taken per semester was 14.3 with standard deviation 0.3.

a. Find a 95% confidence interval for the average number of credit hours Freshmen take per semester.

b. What assumption must be true in order for the above CI to be valid? How is the assumption verified? (Be specific!)

➡ Section 6: Estimation and Confidence Interval Overview:

Inferential Statistics ———➤   Drawing conclusions about the _____ based

on the _____ .

i.e., Using the _____ as an estimate for the

_____ .

Estimation ———➤   estimating the value of a _____ .

Point Estimate: Using one point (one value) to estimate the _____ .

_____ is a point estimate for μ.

_____ is a point estimate for p.

Confidence Interval ———➤   an interval estimate for the _____ where

we include how confident we are that the interval contains

the _____ we are attempting to estimate.

General form of a CI: pt. est. $\pm E$ ———➤ ($pt.est. - E$, $pt. est. + E$)

128   Introduction to Statistics

CI on p:

CI on μ:

➡ Chapter 5 Review Homework:

1. Suppose researchers are interested in estimating the number of hours Freshmen study/do homework per week for their classes. A random sample of 40 Freshmen was obtained. The average hours spent per week studying/doing homework for their classes was 8.3 hours with standard deviation 2.1. Estimate with 95% confidence the true average number of hours Freshmen spend studying/doing homework per week.

2. _____ Suppose the point estimate for the true proportion is 0.40. Further suppose the Margin of Error equals 0.15. Choose the choice below that gives the correct confidence interval.

   A. $0.25 < \hat{p} < 0.55$

   B. $0.325 < \hat{p} < 0.475$

   C. $0.10 < \hat{p} < 0.70$

   D. $\hat{p} = 0.40$

   E. $0.25 < p < 0.55$

   F. $0.325 < p < 0.475$

   G. $0.10 < p < 0.70$

   H. $p = 0.40$

3. Does exercise lower a person's resting pulse rate? This is a question researchers are interested in investigating. Let X be the pulse rate of fifteen randomly selected individuals who are regular attending members of a local gym. The mean of those fifteen pulse rates is 69.95 with standard deviation 7.67.

   a. State the parameter the researcher is interested in estimating? What does the parameter represent in this problem?

   b. Find *and* interpret the margin of error for estimating the parameter of interest with 90% confidence.

   c. Find *and* interpret a 90% confidence interval for the parameter of interest.

4. In a survey of 1219 US adults, 354 said that their favorite sport to watch on TV is women's college basketball.

   a. Estimate with 99% confidence the proportion of US adults who say that their favorite sport to watch on TV is women's college basketball.

   b. Using the answer you found in part b, can you be 99% confident that more than 25% of US adults' favorite sport to watch on TV is women's college basketball? Explain your answer.

   c. ESPN's CEO wants to determine the true proportion of US adults whose favorite sport to watch on TV is women's college basketball to be within 1% accuracy and with 95% confidence. What sample size should they take?

Estimation and Confidence Intervals for One Parameter **131**

5. The Little League Association is interested in determining the average distance a male T-ball player can throw a T-ball. The data below represents the distance thrown in yards for twenty randomly selected male T-ball players.

12	21	20	23	17	12	15	16	14	10
13	23	12	8	18	17	20	19	6	18

a. Estimate with 95% confidence the average distance a male T-ball player can throw a T-ball.

b. The Little League Association has claimed in the past that male T-ball players can throw a T-ball farther than fifteen yards. At 95% confidence, is their claim correct?

HYPOTHESIS TESTING FOR ONE PARAMETER

Statistical Inference – the process of making judgments about a population based on properties of a sample randomly drawn from the population.

Estimation – involves approximating the value of an unknown parameter.

Hypothesis Testing – involves choosing between two opposing statements concerning a parameter.

▶ Section 1: What Is Hypothesis Testing? An Example to Walk Through:

We already know all the definitions and tools we need to know to perform a hypothesis test. We just do not know the formal name of those definitions and tools. Let's walk through an example illustrating how to perform a hypothesis test using the "language" we already know. Then we will go back and give the "formal name".

According to M&M Mars Inc., the proportion of blue M&M's in their snack size pack is 10%. Suppose Ana really loves blue M&M's and always counts how many blue M&M's there are in every snack pack of M&M's she eats. (Ana's mom rewards her daughter's good behavior by giving her a snack pack of M&M's. Since Ana is a good girl, she eats many M&M's.) After many packages of M&M's, Ana begins complaining that there are not enough blue M&M's. So upon investigating this and keeping track of the number of blue M&M's, Ana and her mom discover that out of 240 M&M's, only sixteen were blue. At the 5% significance level can you conclude that the proportion of blue M&M's is less than 10%?

- What is the parameter of interest?

    Answer:

- What does M&M Mars Inc. claim to be true about this parameter?

    Answer:

    Formal Name: _____ .

Hypothesis Testing for One Parameter

- What does Ana think is true about this parameter?

  Answer:

  Formal Name: _____ .

- If I want to make a decision about this parameter _____ , what would I calculate?

  Answer:

- When performing a hypothesis test, what we are trying to decide is, if the true value of the parameter, _____ , is really equal to _____ , is it likely or unlikely for us to calculate a value of _____ = _____ .

- What can we look at to tell us if this value calculated, _____ is a likely or unlikely value if _____ is really equal to _____?

  Answer:

- What is the _____ of _____ ?

  Answer:

- What must be true in order for this to be the _____ of _____?

  Answer:

  Formal Name: _____.

- If _____ is really _____ , which value, A or B, labeled below is most likely to be obtained?

  Why? Answer:

134   Introduction to Statistics

- What do we need to calculate to know where on this curve our _____ value falls?
  Answer:

  Formal Name: _____ .

- What is the distribution?

  Formal Name: _____ .

- Let's calculate its value:

  Formal Name: _____ .

- Is that value a surprising value?
  Answer:

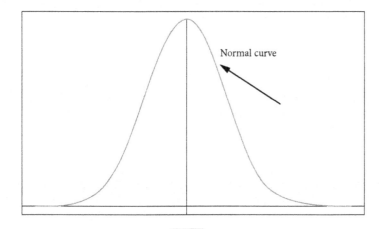

- There are two ways for us to determine if this is a surprising value:
    1. Find the cut-off point such that any value in that range is a "surprising" value: These are the values that support Ana's claim.

    What values support Ana's claim?

0.10

Formal Name: _____ .

- How do we interpret this cut-off point?

Answer:

This says that any _____ value that results in a z-score

_____ to _____ is a _____ value that

shows support for Ana's claim.

Our _____ resulted in the z-score _____ , so our

_____ value _____ Ana's claim.

136 Introduction to Statistics

2. Find the _____ of getting a _____

when _____ is really _____ .

0.10

Formal Name: _____ .

- How do we interpret this _____ ?

  Answer:

  This implies that there is a _____ _____ of observing a

  _____ if _____ is really _____ .

  Is this a small _____ ?

  Answer: _____ , since less than _____ .

- Is _____ , a likely or unlikely value if _____ is really

  _____ ?

  Answer:

- Thus, what does this "evidence" suggest about the value of our parameter? Is there "enough" evidence to support Ana's claim?

  Answer:

Let's now put a "definition name" to each step above. Then we will formally define the "steps" below.

Hypothesis Testing for One Parameter   137

➡ Section 2: Definitions and Steps:

Definitions

Hypothesis: a statement about one or more populations. Usually concerned about the value of a parameter.

Research hypothesis: the conjecture that motivates the research.

Statistical hypothesis: hypotheses that are stated in such a way that they may be evaluated by appropriate statistical techniques.

- The *null hypothesis* – Ho
  - the hypothesis being tested
  - we always assume throughout the test that the null hypothesis is true
  - a statement of agreement with conditions presumed to be true in the population
  - we either reject Ho or fail to reject Ho. **We never accept Ho!**
- The *alternative hypothesis* – Ha
  - corresponds to what the researcher is trying to prove
- <u>Valid Hypothesis:</u> For the null and alternative hypothesis to be valid, it must be of the form:

Ho: parameter $= c$ versus   Ha: parameter $\neq c$
                                  or   parameter $> c$
                                  or   parameter $< c$

To verify that Ho and Ha are valid null and alternative hypothesis, verify the following:

1. Ho and Ha are testing a parameter; not a statistic.
2. Ho is a statement of equality; not an inequality.
3. The value c is the same for both Ho and Ha.
4. The value c is a valid value of the parameter.

- <u>Types of Hypothesis:</u>

The type of test depends on which values show support for the alternative hypothesis. If the values in both tails support Ha, it is a two-tailed test. If large values support Ha (i.e., values in the right tail), it is a right-tailed test. If small values support Ha (i.e., values in the left tail), it is a left tail test.

Two - Tailed	*Right - Tailed*	*Left - Tailed*
Ho: $\mu = 50$	Ho: $\mu = 50$	Ho: $\mu = 50$
Ha: $\mu \neq 50$	Ha: $\mu > 50$	Ha: $\mu < 50$

values in both tails support Ha

values in the right tail support Ha

values in the left tail support Ha

Level of Significance – α: The probability that the test statistic will fall in the rejection region (also called critical region) if the null hypothesis is true is the *level of significance* of the test, which we denote as α. This is set by the researcher. Another way to think about α: It is how the researcher defines "a small probability." The values of α are usually 1%, 5%, or 10%. If the value is not stated we assume α equals 5%.

Test Statistic: The test statistic is the formula used to find the observed value, OV, of the test statistic, where the OV is a quantity that is used to make a decision in a test of hypothesis. The OV tells us where on the distribution curve the point estimate falls.

- General Form of the Test Statistic:

$$T.S. = \frac{\text{relevant statistic} - \text{hypothesized value}}{\text{standard error}}$$

Assumptions of the Test Statistic: The *assumptions of the test statistic* are properties that must be satisfied in order for your test statistic to be valid. The assumptions are what need to be true so that your test statistic will have the assumed distribution.

Rejection Region (RR): The *rejection region* (also called critical region) consists of those values of the test statistic that provide strong evidence in favor of the alternative hypothesis.

Recall, the RR describes for us the "region" on the distribution curve that "unlikely" values of the point estimate will fall if the assumed value of the parameter is true. Thus:

To find rejection region, if Ha:

parameter $< c$ or   parameter $> c$ or   parameter $\neq c$

RR: $Z \leq -z_0$ $Z \geq z_0$ $Z \leq -z_0$ or $Z \geq z_0$

If the OV falls in the RR, we reject Ho. Otherwise, we fail to reject Ho.

P-value: The *p-value* is the probability of getting a value of the point estimate that is favorable or more favorable to the alternative hypothesis, if the null hypothesis is true.

The p-value describes for us, using probability, if the point estimate is an "unlikely" or "likely" value, if the assumed value of the parameter is true.

To find p-value, if Ha:

	parameter $< c$	parameter $\neq c$	parameter $> c$
p-value:	(for ex:) $\Pr(\hat{p} \leq \underline{\hspace{1cm}}) = ?$	$2\Pr(\hat{p} \leq \underline{\hspace{1cm}}) = ?$ (if OV is negative) $2\Pr(\hat{p} \geq \underline{\hspace{1cm}}) = ?$ (if OV is positive)	$\Pr(\hat{p} \geq \underline{\hspace{1cm}}) = ?$

Small probabilities imply "unlikely," where "small" is defined as "less than α." Thus:

If the p-value is less than α, we reject Ho. Otherwise we fail to reject Ho.

Decision of the test: The *decision* of the hypothesis test is either reject Ho or fail to reject Ho.

To determine the decision of the test, we ask ourselves two questions:

1. Is the OV in the RR? If we say yes, we reject Ho. If we say no, we fail to reject Ho.

2. Is the p-value < α? If we say yes, we reject Ho. If we say no, we fail to reject Ho.

If Ho is rejected, then we no longer know the value of the parameter. Thus, we want to estimate its value using a confidence interval. When using a CI to estimate a parameter from a hypothesis test, the CI needs to correspond to the type of hypothesis test constructed. If a one-tailed hypothesis test is performed, a one-bound CI will be found with the appropriate bound being determined by the type of test. If a left-tail test is performed, only the upper bound is calculated. If a right tail test is executed, only the lower bound is obtained. If a two-tailed hypothesis test is performed, then both the lower bound and the upper bound of the CI are calculated. The table below illustrates this.

To find the CI, if Ha:

parameter $< c$	parameter $\neq c$	parameter $> c$
pt.est. + |RF|*(SE) (using α)	pt.est. ± |RF|*(SE) (using α/2)	pt.est. – |RF|*(SE) (using α)

There are many "steps" involved in hypothesis testing. In order to have your thoughts and work organized so that you can understand and complete hypothesis testing, you will follow the following steps when working a hypothesis problem. The steps are geared to walk you through the thought process of hypothesis testing.

Steps for hypothesis testing: _ _ _ _ _ _ _ _ _ _ _ _

1. Define population and parameter of interest.

2. State α = the level of significance.

3. Set up null and alternative hypothesis:

4. State the test statistic and give its distribution. (The distribution comes from the sampling distribution of the statistic.)

5. State and verify assumptions.

6. Analyze the data. In other words, state the point estimate and find the observed value of the test statistic.

7. Determine if the evidence is significant. In other words, find the rejection region and the p-value.

8. Conclusion – State your decision and your conclusion in terms of the problem.

9. If Ho is rejected, find and interpret the appropriate Confidence Interval.

140   Introduction to Statistics

Checking your Understanding: (Answers are found in the Appendix)

1. Suppose the observed value equals 2.34 and the rejection region is $Z_p \geq 1.645$. What is the decision of the test?

2. Suppose the p-value of the test equals 0.0215 and the level of significance equals 5%. What is the decision of the test?

3. Suppose the following null and alternative hypothesis is being tested: Ho: $p = 0.5$ versus Ha: $p > 0.5$. Further suppose the decision of the test is Reject Ho. Which bound of the confidence interval should be constructed?

Section 2: Homework

Fill in the blank with the correct term/definition:

1. The _____ consists of those values of the test statistic that provide strong evidence in favor of the alternative hypothesis.

2. The _____ is the probability of getting a value of the test statistic as favorable or more favorable to the alternative hypothesis than the observed value, assuming the null hypothesis is true.

3. The probability that the test statistic will fall in the rejection region (critical region) if the null hypothesis is true is the _____ of the test, which is denoted as _____.

4. The _____ is a quantity that is used to make a decision in a test of hypothesis.

5. The _____ are properties that must be satisfied in order for your test statistic to be valid.

6. The _____ hypothesis corresponds to what the researcher is attempting to prove.

7. We always assume throughout the test that the _____ hypothesis is true.

➡ Section 3: Hypothesis Test for One Population Proportion, p:

For the population proportion,

Parameter:

Point Estimate:

Hypothesis Testing for One Parameter **141**

Test Statistic:

Assumption:

Example 1: Consider the case of a software company that developed a general theme for a new game based on the assumption that half of its customers are over twenty years of age. The agency wants to change the general theme of this new game if this proportion has decreased. The agency conducted a survey of a sample of 400 of its customers. Of the 400 customers, 190 were over the age of twenty and 210 were twenty years or less. At the 5% significance level, what conclusion can be reached?

1. Population: Parameter of interest:

2. $\alpha =$

3. Set up hypothesis: Ho:

                        Ha:

4. State the test statistic and give its distribution.

5. State and check assumptions.

6. Analyze data.
   Point Estimate:

   ┌───┐
   │ Idea Behind Hypothesis Testing: │
   │   │
   │ Recall, we are trying to determine, if   │
   │   │
   │ _____ are we surprised or not │
   │   │
   │ surprised to get _____. │
   └───┘

142   Introduction to Statistics

Observed Value:

What is step 6 describing for us?

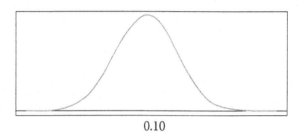
0.10

7. Determine if the evidence is significant. (i.e., find the rejection region and the p-value)
rejection region:

0.10

p-value:

0.10

Hypothesis Testing for One Parameter **143**

8. Decision:

Conclusion:

9. If Ho is rejected, find and interpret the appropriate Confidence Interval.

Another Example:

Pediatric researchers sampled 680 infants and found that 68% of them have had the MMR vaccine. Can we conclude on the basis of these data that the percent of all infants that have had the MMR vaccine is different than 62%? Assume that the researcher has set the level of significance to be 2.5%.

Checking your Understanding: (Answers are found in the Appendix)

The Southern Online Viewing (SOW) Company claims that 60% of online video subscribers recommend its viewing platform over that of the Northern Online Viewing (NOW) Company's platform. To test this claim against the alternative that the actual proportion is less than 60%, a random sample of 50 online viewers was taken and 52% said they prefer SOW's platform over NOW's. At the 5% significance level, what conclusion can be reached?

Section 3: Homework

Show all work/steps!

1. The La Leche Society has released an educational video on the benefits of breast feeding. Prior to the release of the videos, it was known that 75% of new moms breast-feed exclusively for the first six weeks of their child's life. The CEO wants to know if the videos are increasing this proportion. A random sample of 400 new moms was obtained and 320 breast-fed their new babies exclusively for the first six weeks. At the 5% significance level, what conclusion can be reached?

2. Although many new moms will nurse their new babies in the beginning, the Pediatrics Association believes that 50% of moms are still nursing their babies by the time the babies are six months old. The La Leche Society believes that this proportion is less than 50% and thus believes more educational information should be given to new moms to encourage the mothers to nurse for the first year of their child's life. A random sample of 425 new moms was obtained and 200 were still nursing their child at six months of age. At the 2.5% significance level, what conclusion can be reached?

3. With the increase in childhood obesity, the Pediatrics Association is concerned about the amount of physical activity children obtain daily. In 2012 the Pediatrics Association found that only 48% of children between the ages of five and twelve were physically active for at least one hour a day, where physically active refers to any activity that increases the child's heart rate to the aerobic zone. A random sample of 250 children between the ages of five and twelve was obtained. Of those sampled, 59% said they were physically active for at least one hour a day. At the 1% significance level, can the Pediatrics Association conclude that the proportion of children between the ages of five and twelve that are physically active for at least one hour a day has changed.

4. Many people assume that nurses are less likely than others to "feel faint" or "get sick" at the sight of blood. Suppose it is known that 30% of Americans will feel faint or get sick at the sight of blood. A random sample of 500 nurses was obtained. Of the 500 nurses, only 125 admitted to ever feeling faint or getting sick at the sight of blood. At the 10% significance level, can you conclude that nurses are less likely than other Americans to feel faint or get sick at the sight of blood?

➡️ Section 4: Type of Errors and Power:

Types of Errors: Every time we make a decision (either reject Ho or fail to reject Ho), we either made the correct decision or we made an error. There are two types of errors:

- *Type-I error*: The *Type-I error* is rejecting a true null hypothesis. We denote the probability of making a Type-I error as α.

- *Type-II error*: The *Type-II error* is failing to reject a false null hypothesis. We denote the probability of making a Type-II error as β.

- *Power*: The *power* of the test is the probability of rejecting a false null hypothesis.
  Power = $1 - \beta$.

- More on α and β:

  1. α = probability of type I error = probability of getting a false positive = level of significance
     α is known! (Set by the researcher)

  2. β = probability of type II error = probability of getting a false negative
     β is unknown!

  3. If we increase α, we decrease β and vice versa.

146 Introduction to Statistics

4. Note that $\alpha + \beta \neq 1$, since there are more than just two possibilities when performing a hypothesis test: making a Type I error, making a Type II error, or making the correct decision.

○ Example: (Court of Law Example)

Ho: innocent

Ha: guilty

		Reality	
		Innocent *Ho is true*	Guilty *Ho is false*
Jury Says	Innocent *Ho is true* (fail to reject Ho)	Correct Action True Negative Probability = $1 - \alpha$	Type II Error False Negative Probability = β
	Guilty *Ho is false* (reject Ho)	Type I Error False Positive Probability = α	Power True Positive Probability = $1 - \beta$

- If we reject Ho, then we possibly made what type of error?

  Answer:

- If we fail to reject Ho, then we possibly made what type of error?

  Answer:

- Why is it that we never accept the Ho:

  Answer:

- Why is it that we do reject Ho:

  Answer:

- Suppose you have the following hypothesis: Ho: House is on fire

  Ha: House is not on fire

In the context of the null and alternative hypothesis, what is the Type-1 error?

Answer:

In the context of the null and alternative hypothesis, what is the Type-2 error?

Answer:

In terms of the consequences of making the error, which of the above is the "worse" error?

Answer:

In hypothesis testing, when constructing the null and alternative hypothesis, they should be constructed in such a way that the "worse" error corresponds to a Type-_____ error.

Why?

- When choosing the test statistic, we want to choose the test statistic that has what two properties?
  1.

  2.

Power is increased/maximized by:

1. Choosing the correct method and verifying assumptions.

2. Increasing the distance between the true value of the parameter and the hypothesized value of the parameter.

3. Decreasing the variability in the data.

4. Increasing α.

5. Increasing n.

Our goal is to have _____ power while controlling for _____ .

148   Introduction to Statistics

- More on # 1: In every testing situation, there is a "gold standard" test. The test that has maximum power and controls for alpha. It is also the case that every test has assumptions that must be true. If those assumptions are not valid, then the test loses its power. It is of upmost importance that one always verifies the validity of the assumptions before performing the test.

- More on #2 and #3: As a researcher we cannot control the distance between the true value of the parameter and the hypothesized value of the parameter (what the null hypothesis claims) nor can we control for how much variability there is in the data. However, a researcher should understand how this affects the power of the test.

- More on #4: A researcher can increase α to increase power. However, there is clearly a trade-off. Furthermore we are limited to how large we can increase α. Recall the largest acceptable α in research is 10% however most scientific research is performed at the 5% significance level.

Checking your Understanding: (Answers are found in the Appendix)

1. Suppose you have the following null and alternative hypothesis:

   Ho: It is too cold to ski

   Ha: It is not too cold to ski

   a. In the context of this problem, describe the power of the test.

   b. In the context of this problem, describe what α represents.

   c. In the context of this problem, describe what β represents.

2. To increase power, should n be increased or decreased?

3. To increase power, should α be increased or decreased?

4. Suppose the critical value of a two-tailed test equals $|2.064|$.

   a. What is the rejection region?

   b. Suppose the observed value equals -3.45. What is the decision of the test?

5. Suppose the p-value equals 0.0447 and the level of significance equals 5%. What is the decision of the test?

Section 4: Homework

1. Suppose you have the following null and alternative hypothesis: Ho: It is snowing

   Ha: It is not snowing

   a. In the context of this problem, describe a Type-I error.

   b. In the context of this problem, describe a Type-II error.

   c. In the context of this problem, describe the power of the test

2. Suppose the observed value of the test equals 2.45 and the critical value of the rejection region for a right tailed test equals 2.281. What is the decision of the test?

3. Suppose the observed value of the test equals -1.95 and the critical value of the rejection region for a left tailed test equals -2.371. What is the decision of the test?

Hypothesis Testing for One Parameter **149**

4. Suppose the p-value for a two-tailed test equals 0.0463 and the level of significance equals 2.5%. What is the decision of the test?

5. Suppose the p-value for a two-tailed test equals 0.0241 and the level of significance equals 2.5%. What is the decision of the test?

6. What is the effect on the power of the test for the following scenarios:

   a. β is increased

   b. β is decreased

   c. α is increased

   d. α is decreased

   e. n is increased

   f. n is decreased

7. What is the effect on β for the following scenarios:

   a. n is increased

   b. n is decreased

   c. α is increased

   d. α is decreased

8. Suppose your roommate believes the mean amount it will cost to repair your dorm refrigerator is more than $50. To determine if she is correct, suppose you perform a hypothesis test with level of significance equal to 5%. For each choice below, determine if you made the correct decision or if you made an error. If an error was made, determine if a Type-I or Type-II error was committed.

   a. You conclude that the mean amount to repair a dorm refrigerator is more than $50 when in reality the mean amount to repair a dorm refrigerator is not more than $50.

   b. You conclude that the mean amount to repair a dorm refrigerator is not more than $50 when in reality the mean amount to repair a dorm refrigerator is not more than $50.

   c. You conclude that the mean amount to repair a dorm refrigerator is not more than $50 when in reality the mean amount to repair a dorm refrigerator is more than $50.

   d. You conclude that the mean amount to repair a dorm refrigerator is more than $50 when in reality the mean amount to repair a dorm refrigerator is more than $50.

9. For the each of the null and alternative hypothesis below, determine if it represents a valid null and alternative hypothesis. If the null or alternative hypothesis is not valid, explain why.

   a. Ho: $\mu = 20$
   Ha: $\mu \geq 20$

   b. Ho: $\mu \neq 18$
   Ha: $\mu = 18$

   c. Ho: $\mu = 0$
   Ha: $\mu < 0$

   d. Ho: $\bar{x} = 15$
   Ha: $\bar{x} > 15$

   e. Ho: $\mu = 20$
   Ha: $\mu > 19$

   f. Ho: $p = 5$
   Ha: $p > 5$

   g. Ho: $p < 0.20$
   Ha: $p > 0.20$

   h. Ho: $p = \frac{1}{2}$
   Ha: $p \neq \frac{1}{2}$

150   Introduction to Statistics

10. Fill in the blank with the correct definition term or symbol.

    a. A _____ is failing to reject a false null hypothesis.

    b. A _____ is rejecting a true null hypothesis.

    c. The probability of rejecting a false null hypothesis is called the _____ of the test.

    d. The probability of making a Type-II error is denoted as _____ .

    e. The probability of making a Type-I error is denoted as _____ .

➡ Section 5: Hypothesis Test for μ When σ Is Unknown:

Recall the general form of the test statistic:

$$\text{T.S.} = \frac{\text{relevant statistic} - \text{hypothesized value}}{\text{standard error}}$$

Thus, for the population mean, the test statistic is:

Answer:

But once again, is it likely that we know σ?   So, we have for our formula:

Answer:

Thus we have for testing one population mean with the population standard deviation unknown:

Parameter:

Point Estimate:

Test Statistic:

Assumption:

Example 1: A manufacturer of motorcycle exhaust systems wants to test its new exhaust system to determine whether it meets state air pollution standards. The mean emission of all motorcycle exhaust systems of this type must be less than twenty parts per million of carbon. Fourteen exhaust systems were randomly selected and tested. The mean emission level was found to be seventeen ppm with standard deviation 2.94. Do the data supply sufficient evidence to allow the manufacturer to conclude that the new exhaust system for motorcycles meets the pollution standard? Assume that the manufacturer has set the probability of making a Type-I error to 1%. Show all steps.

- If the data is given, we can let JMP do the work for us. Below is how the problem will look if data is given. The only steps that change are steps 5, 6, part of 7, and if needed step 9. You still need to show ALL the steps! JMP will just simply do the calculations for us on the steps mentioned.

Example 2: A manufacturer of motorcycle exhaust systems wants to test its new exhaust system to determine whether it meets state air pollution standards. The mean emission of all motorcycle exhaust systems of this type must be less than twenty parts per million of carbon. Fourteen exhaust systems are manufactured for testing purposes. The data below represents the emission level of the fourteen exhaust systems tested. Do the data supply sufficient evidence to allow the manufacturer

Hypothesis Testing for One Parameter   **153**

to conclude that the new exhaust system for motorcycles meets the pollution standard? Assume that the manufacturer has set the level of significance to 1%.

17	16	20	18	16	19	22
13	15	17	21	14	18	12

Show which steps change:

154   Introduction to Statistics

JMP Instructions:

1. Input data
2. Under analyze, choose *distribution*. Click on the column containing the data, click on *Y, Columns*, and click on *Ok*.
3. Under the red options button next to distribution, choose *stack*
4. Under red options button next to column 1 (or whatever you named your column), choose *normal quantile plot* (to check normality).
5. Under red options button next to column 1 (or whatever you named your column), choose *test mean*.
6. Type in the hypothesized value of the mean, hit okay.
7. If need CI, under red options button next to column 1 (or whatever you named your column), choose *confidence interval*, the choose *other* to obtain lower bound only or upper bound only. Make sure and enter the confidence level.

- JMP Output:

Distributions
Column 1

Test Mean	
Hypothesized Value	20
Actual Estimate	17
DF	13
Std Dev	2.9352

Hypothesis Testing for One Parameter 155

	t Test
Test Statistic	-3.8243
Prob > |t|	0.0021*
Prob > t	0.9989
Prob < t	0.0011*

One-sided Confidence Interval				
Parameter	**Estimate**	**Lower CI**	**Upper CI**	**1-Alpha**
Mean	17	.	19.07907	0.990
Std Dev	2.935198	.	5.222171	0.990

Checking your Understanding: (Answers are found in the Appendix)

1. The Southern Online Viewing (SOW) Company would like to see if they are increasing their average length of viewing per customer over last year. The average amount of SOW content watched by a subscriber per day last year was 1.8 hours. A random sample of 68 viewers found that the average length of viewership was 1.85 hours with a standard deviation of 0.45 hours. At the 5% significance level, is there significant evidence to conclude the claim from SOW? Assume all necessary assumptions are valid.

2. Suppose a researcher in public health is interested in determining if the average starting salary for those with a master's degree in public health is changing. The data below represents the starting salaries (in thousands) for 22 recent graduates that received a master's degree in public health. At the 1% significance level, can the researcher conclude that the average starting salary for those with a master's degree in public health is different from $55,000?

66.251 67.441 53.159 56.017 61.264 64.281 61.488 63.665 66.792 63.107 53.636

55.236 54.651 61.213 58.26 55.28 57.421 55.102 44.188 49.53 45.767 60.564

Section 5: Homework

Show all work/steps! For the problems containing data, you must use JMP to analyze the data.

1. Are college students spending more or less time studying for class than in the past? This was a question of interest for researchers. Twenty years ago, college students spent an average of 2.8 hours a day studying for class. A random sample of fifty present day college students was obtained and the average time they spend studying for class was found to be 2.4 hours with standard deviation 1.4. At the 2% significance level can the researchers conclude that the average amount of time college students spend studying for class is different than the average amount of time in the past spent studying? Assume the time spent studying is a normally distributed random variable.

2. Is the strep virus getting harder or easier to heal? Pediatricians are interested in studying the average length of time patients with strep are running fever while on antibiotics. In previous years, the average length of time patients ran fever with strep while on antibiotics was 1.8 days. A random sample of forty-six recent patients with strep ran fever while on antibiotics for an average of 2.3 days with standard deviation 1.2. At the 5% significance level, can pediatricians conclude that the average amount of time patients with strep run fever while on antibiotics has changed? Assume the random variable has a normal distribution.

156   Introduction to Statistics

3. Dieticians claim that most adults do not intake enough dietary fiber from food (not supplements). For optimal health, adults should intake an average of twenty-eight grams of fiber from food per day. Below is the amount of dietary fiber from food consumed by twenty randomly selected adults. At the 5% significance level, can we conclude that the average amount of dietary fiber from food consumed per day is less than twenty-eight grams?

22	23	18	15	17	14	25	20	21	29
28	14	12	15	18	13	17	21	24	19

4. Is the flu making us sicker? A question of interest at the latest health summit was whether the latest strain of flu was causing adults to run a higher fever than previous strains. The average temperature for previous strains of flu was 101.4 degrees. A random sample of eighteen adults with the latest strain of flu was obtained. The data below represents their highest temperature while sick with the latest flu strain. Based on the data, can we conclude that the latest strain of flu is causing patients to run a higher fever, on average, than previous strains of flu? Let the probability of making a Type-I error be 10%.

102.1	101.8	100.9	103.8	102.8	100.8	102.5	102.0	103.1
104.0	104.2	101.1	100.3	101.9	101.7	101.8	102.3	104.1

➡️ Section 6: The Relationship between Confidence Intervals and Hypothesis Testing:

Recall when finding a CI, it was referred to as a $(1 - \alpha)$% CI. Thus for a 90% CI, $\alpha = 10$%.

Where else have we seen α? It is the probability of making a Type-_____ error. Hence, there is a relationship between CI and hypothesis testing. A CI can be used to make a decision in a hypothesis test.

Recall that when we discussed CI, in order for a CI to support a claim, _____ value in the CI has to support the claim.

Thus, to use a CI to draw a conclusion in a hypothesis test, we look to see if every value in the CI supports the alternative hypothesis. Another way to think of it is: if the null hypothesis could be true, we cannot reject it. Thus, if the value of the null hypothesis is contained within the CI, we must fail to reject Ho.

- Example 1: What percent of college students buy a pet? This is a question of concern for the local humane society. The local humane society would like to estimate with 95% confidence the proportion of college students that buy a pet. A random sample of 200 college students was obtained. Of those sampled, 44% had purchased a pet.

  - A 95% CI for _____:

Hypothesis Testing for One Parameter   **157**

- The humane society would like to know if the true proportion of college students that purchase a pet is different than 50%. This gives us the following null and alternative hypothesis:

  Ho:

  Ha:

- Based on the CI, what is the decision and conclusion of the hypothesis test?

Checking your Understanding: (Answers are found in the Appendix)

An educational researcher wants to determine if the percentage of high school students who like math is different from 50%. She takes a random sample of 85 high school students and finds that 32 of them like math.

a. What is the null and alternative hypothesis that corresponds to this problem?

b. Find a 95% confidence interval for the proportion of high school students who like math.

c. Based on the confidence interval found in part b, what would be the decision of the hypothesis test for part a?

Section 6: Homework

1. With economical concerns being so great, many people are no longer purchasing new vehicles, but are instead keeping their vehicles for a longer period of time, until they no longer work. The American Automobile Association is interested in determining if the average age of vehicles for Americans is different than six years. A random sample of 35 Americans was obtained. The average age of their vehicles was 6.2 years with standard deviation 1.2.

   a. Set up the null and alternative hypothesis:

   b. At the 1% significance level, use a CI to determine if the average age of vehicles for Americans is different than six years. Make sure and give your CI and your decision and conclusion.

2. Suppose a 95% CI for the proportion of adults that regularly attend a tae kwon do class at least once a week was found to be (0.2534, 0.3894).

   a. For the null and alternative hypothesis: Ho: $p = 0.40$ versus Ha: $p \neq 0.40$, what is your decision for this test of hypothesis?

158   Introduction to Statistics

b. Based on the CI, which of the following statements gives the best conclusion about the true value of the proportion of adults that regularly attend a tae kwon do class at least once a week?

    i. I am 95% confident that the proportion of adults that regularly attend a tae kwon do class at least once a week is greater than 40%.

    ii. I am 95% confident that the proportion of adults that regularly attend a tae kwon do class at least once a week is different than 40%

    iii. I am 95% confident that the proportion of adults that regularly attend a tae kwon do class at least once a week is less than 40%

    iv. I am 95% confident that the proportion of adults that regularly attend a tae kwon do class at least once a week equals 40%

Section 7: Hypothesis Testing for One Parameter Overview:

Fill in the following for testing one population proportion:

Parameter:

Point Estimate:

General form for Ho:

        Ha:

Test Statistic and its distribution:

Assumptions:

Hypothesis Testing for One Parameter   **159**

General form of CI:

Fill in the following for testing one population mean:

Parameter:

Point Estimate:

General form for Ho:

Ha:

Test Statistic and its distribution:

Assumptions:

General form of CI:

General form of the RR (for both parameters):

General form of probability statement for the p-value (for both parameters):

➡ Chapter 6 Review Homework:

1. In a random sample of 28 four-person tents, the mean price was $344.19 and the standard deviation was $61.32. Assume the prices are normally distributed. At the 5% significance level can we conclude that the mean price of four-person tents has increased from the previous mean price of $325?

2. A hot topic of debate is whether the legal driving age should be raised to eighteen. From a random sample of 400 Americans, 52% favored raising the legal driving age to eighteen. Based on the data, can you conclude that a majority of Americans believe the legal driving age should be raised to 18?

   Let $\alpha = 2.5\%$

3. Are our kids getting taller? The average height of kindergartners in 1950 was 39.5 in. The data below represents the height (in inches) of twenty-six randomly selected kindergartners. At the 5% significance level, what conclusion can be reached?

40.1	39.9	41.2	41.3	39.2	39.3	38.7	38.8	39.2	40.3	41.1	40.7	39.8
39.7	40.9	40.0	39.0	41.0	40.6	39.6	39.5	40.5	40.8	40.9	39.2	41.1

4. Due to the economy, are Americans staying in their first professional job longer? This was a question of interest for the Federal Jobs Bureau. In 2010, Americans were staying an average of 2.3 years in their first professional job. The data listed below is the length of time (in years) that eighteen randomly selected adults were staying in their first professional job. At the 5% significance level, what conclusion can be reached?

2.4	2.1	1.9	3.1	3.2	2.9	2.6	1.9	3.8
2.3	2.5	1.8	2.2	4.5	5.1	2.1	2.4	6.1

Hypothesis Testing for One Parameter **163**

CHAPTER SEVEN
HYPOTHESIS TESTING FOR TWO PARAMETERS

Many times we are interested in comparing two populations, thus we are interested in comparing two parameters. For CI and for hypothesis testing, the concepts and general idea are the same. So, to determine the formula for the CI or to determine which hypothesis test to use, we just need to know based on the parameter of interest, what is the point estimate and what is the distribution and standard deviation of the point estimate.

▶ Section 1: Independent vs. Dependent Samples:

When testing *two* population parameters, one criterion in choosing the test statistic depends on whether the samples are independent or dependent. Thus, the first question we should ask ourselves when we realize we are testing two parameters is "Are the samples independent or dependent?" In other words, "Did we obtain the samples in such a way that they are independent or did we obtain them in such a way that they are dependent?"

- Two samples are *independent* if the data values obtained from one sample are unrelated to the values obtained in the second sample.

Population 1   Population 2

- The samples are *dependent* if each data value from one sample is paired in a natural way with a data value from another sample.

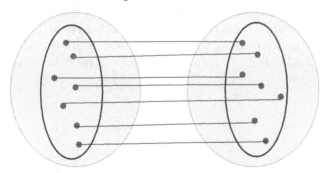

Population 1   Population 2

Hypothesis Testing for Two Parameters **165**

- Why use "paired" or "dependent" samples? Because it can eliminate extraneous or confounding factors.

  For Example: Suppose we want to test two sunscreens—Sunscreen A and Sunscreen B.

  - Experiment 1: We could randomly assign one group to receive Sunscreen A and randomly assign another group to receive Sunscreen B. We then look at the amount of sun damage after a specific amount of time.

    - Is this an example of independent or dependent samples?
      Answer:

    - Is there something other than the brand of sunscreen that can affect the amount of sun damage?
      Answer:

  - Experiment 2: We could randomly choose one group of people and apply Sunscreen A to the right half of their back and Sunscreen B to the left half of their back. We then look at the amount of sun damage after a specific amount of time.

    - Is this an example of independent or dependent samples?
      Answer:

    - Does this control for the extraneous factor?
      Answer:

- How do we obtain "paired" or "dependent" samples?

  1. have pre and post measurements from the same subject

  2. have litter mates of same sex

  3. twins

  4. match people based on some characteristic

Checking Your Understanding: (Answers are found in the Appendix)

Researchers are interested in comparing two different surgical methods for correcting cataracts. A random sample of 20 Americans that have cataracts in both eyes was obtained. Phacoemulsification surgery was performed on one eye and Extracapsular surgery was performed on the other eye of the randomly chosen subject. Researchers were interested in determining which method was most successful in correcting the cataract with minimal pain and side effects. Is this an example of independent or dependent samples?

166   Introduction to Statistics

Section 1: Homework

1. Researchers wanted to compare the efficacy of non-surgical versus surgical methods for curing cataracts. A random sample of 20 Americans with a cataract was obtained. These Americans had Extracapsular surgery performed to treat the cataract. Another random sample of 20 Americans with a cataract was obtained. These Americans used cataract eye drops to treat the cataract. Is this an example of independent or dependent samples?

2. Researchers are interested in determining the success of "non-surgical" methods for curing cataracts. A random sample of 20 Americans was obtained that have a cataract in one eye. The subjects were paired based on the severity of the cataract. Of the pairs, one used herbs to treat the cataracts and the other used eye drops. Is this an example of independent or dependent samples?

▶ Section 2: Hypothesis Testing for Two Population Means with Dependent Samples:

When testing two population means from dependent samples we analyze the *differenced data*.

- For example, suppose we have the following data set from two dependent samples:

X_1	X_2	$X_d = X_1 - X_2$
66	60	6
82	79	3
96	92	4
⋮	⋮	⋮

- We define a new random variable, $X_d = X_1 - X_2$ and a new parameter $\mu_d = \mu_1 - \mu_2$. This takes us from two samples to one sample and hence from testing two parameters to testing one parameter: the population differenced mean, μ_d. So, to proceed, we do as before: if σ is known we perform the z-test: if σ is unknown we perform the t-test.

- Thus, the parameter of interest = _____

- The point estimate = _____

- Null and Alternative hypothesis:
    - If we want to determine if $\mu_1 \neq \mu_2$, then this is equivalent to testing

        Ho: Ha:

    - If we want to determine if $\mu_1 > \mu_2$, then this is equivalent to testing

        Ho: Ha:

Hypothesis Testing for Two Parameters   **167**

- If we want to determine if $\mu_1 < \mu_2$, then this is equivalent to testing

  Ho: Ha:

- Test Statistic and its distribution:

  Answer:

- General form of the CI:

  Answer:

- Assumption:

  Answer:

- Example 1: An educational specialist for kindergarten students was interested in finding out whether a creative approach program would increase the scores of kindergartners on a reading ability test. Sixteen students were selected and paired according to a previous test on letter recognition ability. The scores on the reading ability test are as follows:

Creative Approach Program	Standard Program
76	69
89	91
96	92
82	73
92	85
77	69
85	82
85	86

At the 5% significance level, do the data suggest that the creative approach program is effective in increasing the scores of the kindergartners on the reading ability test?

Does the CI agree with your decision in step 8? Why or Why not?

Answer:

JMP Instructions:

1. Put the data into two separate columns
2. Under *columns*, choose *new column*
3. Name the column *difference*
4. Under *column properties*, choose *formula*
5. Click on column 1
6. Click on minus sign
7. Click on column 2
8. Hit okay
9. Under *Analyze*, choose *distribution*
10. Click on difference column, click on *Y*, hit okay
11. Follow the steps for using a *t*-test as done in Chapter 6.

JMP Output:

Distributions
d in test scores

Test Mean	
Hypothesized Value	0
Actual Estimate	4.375
DF	7
Std Dev	4.13824
t Test	
Test Statistic	2.9903
Prob > |t|	0.0202*
Prob > t	0.0101*
Prob < t	0.9899

One-sided Confidence Interval				
Parameter	Estimate	Lower CI	Upper CI	1-Alpha
Mean	4.375	1.603066		0.950
Std Dev	4.138236	2.919184		0.950

170   Introduction to Statistics

Practical Significance – VS – Statistical Significance:

Based on the CI, do you believe that the Creative Approach program does a "good" job on raising the reading ability for kindergartners?

Answer:

As the superintendent of the school district looking to incorporate the program, knowing that it will cost quite a bit of money to implement the Creative Approach program, would you keep the alpha level set at 5% or would you lower the alpha level to say, 1%?

Answer:

As the superintendent, if you lower the alpha level to 1%, without "redoing" any work, what would your decision be?

Answer:

In interpreting a hypothesis test result, one should always take into account that a test can show a statistically significance difference and yet in reality it could be the case that there is not enough difference to be practically significant! This is why a good statistical report should give the p-value and the CI, since they help us in determining practical significance as well as statistical significance.

Hypothesis Testing for Two Parameters

Checking your Understanding: (Answers are found in the Appendix)

1. An article written in Healthy Choices magazine claimed that you would gain weight once you sub-scribe to an online movie repository site. Data analysts for the Southern Online Viewing (SOW) Company wanted to prove the article wrong, so they surveyed 20 new subscribers and measured their weight. After two years they followed up with them to measure their weight again. The data is given below. At the 1% significance level, do the analysts have evidence to support the article's claim?

Weight Before																			
128	112	124	116	119	127	115	126	121	120	119.5	128	115.5	121	119.5	117.5	120.5	118.5	111	122
Weight After																			
129.5	118	130.5	118	128	133	120.5	131	120	118	126	129.5	124	125	117	122	124.5	116.5	114.5	129

Section 2: Homework

Show all work/steps! You must use JMP to analyze the data.

1. Does the bat make a difference? This was a question of interest at the regional little league monthly meeting. To test whether the more expensive bats do increase the average distance hit, ten little league batters of the same age were randomly chosen. The batters hit ten balls with the top of the line bat and after a rest; hit ten balls with the lower price bat. The data below represents the distance hit in feet for the batters.

   At the 5% significance level, does the evidence suggest that the top of the line bat does increase the average distance hit versus the lower price bat?

Top of the Line Bat	69.9	59.2	58.9	58.7	63.2	62.5	66.5	70.4	65.9	55.5
Lower Price Bat	59.2	61.2	52.6	54.3	58.1	62.3	56.5	62.4	53	61.8

2. Does owning an indoor dog really help bring down borderline high systolic blood pressure? To answer this question, a random sample of twelve subjects with chronically borderline high sys-tolic blood pressure was obtained. Their systolic blood pressure was recorded. They then received a house-trained dog to live indoor with them. After one month, their blood pressure was taken again. At the 2.5% significance level, can you conclude that owning a pet will reduce the average systolic blood pressure?

Blood Pressure Before	139.9	130.4	131.3	130.8	128.9	129.5	130.8	129.3	128.1	133.4	125.2	123.9
Blood Presure After	136.3	127.4	125.6	131.2	127.5	126.9	130.1	125.7	130.4	134.8	125.6	121.9

172   Introduction to Statistics

3. A popular myth is that for identical twins, one is the "nice one" and one is the "mean one." Fourteen pairs of identical twins were randomly selected and given a personality test for measuring "niceness" or "meanness" where the higher score implies "niceness" and a lower score implies "meanness." At the 1% significance level, is there a difference in the personality scores between the identical twins.

First Born Score	43	54	49	53	49	52	58	51	59	48	45	49	47	46
Second Born Score	52	47	50	46	50	51	50	45	46	47	49	54	50	54

➡ Section 3: Hypothesis Testing for Two Population Means with Independent Samples:

When testing two population means from independent samples, we cannot analyze the differenced data like we did for dependent samples. We must keep the data sets separate and choose the appropriate test; the test that controls for alpha and maximizes power. Once again the main idea for hypothesis testing and CI remain the same. We just need to know the following:

- The parameter of interest = _____

- The point estimate = _____

- General form for Null and Alternative hypothesis:
  Answer:

- Test Statistic and its distribution:
  Answer:

$$t_p = \frac{(\bar{x}_2 - \bar{x}_1) - (\mu_2 - \mu_1)}{S_P \sqrt{\frac{1}{n_1} + \frac{1}{n_2}}} \ \dot{\sim}\ t_{(n_1 + n_2 - 2)}$$

- General form of the CI:
  Answer:

$$(\bar{x}_2 - \bar{x}_1) \pm t \cdot S_P \sqrt{\frac{1}{n_1} + \frac{1}{n_2}}$$

If performing a two-tailed test, JMP will calculate the CI. If, however, we are performing a one-tailed test, we must use the above formula to find the CI.

- Assumptions:
  Answer:

- Thus, when using the pooled *t*-test, we have **two** assumptions that we have to check (in step 5):

  1. Verifying normality:

  2. Verifying equal population variances: We will verify this using JMP, where JMP is doing a hypothesis test for the null and alternative hypothesis.

     Ho: vs.   Ha:

     - Do we want to   reject Ho   or   fail to reject Ho?

     - To make our decision, JMP will provide us a p-value. Do we want a "large"   or   "small"   p-value?

     - How do we define a _____ p-value?
     - Are we more concerned here with a Type-1 or Type-2 error?
       - What is a Type-1 error here?

     Consequence:

       - What is a Type-2 error here?

174   Introduction to Statistics

Consequence:

- Which is "worse"?

  So, are we more concerned with α or β?

  So to minimize β, we should _____ α.

- Thus, we will define a _____ p-value as p-value _____ .

- Example: Do students taking Introduction to Statistics in early morning classes perform better on their final exam than their peers in late afternoon classes? The following are the student's final exam scores. Using a 5% level of significance, does the early morning class seem to perform better on the final exam than the late afternoon class?

Early Morning Class:	80	82	71	90	69	75	78	87	
	95	98	86	79	77	84	81	91	
Late Afternoon Class:	79	83	73	77	84	90	91	95	80
	94	75	76	69	89	80	82	81	

Do we need to perform step 9? Regardless, let's perform step 9:

Does the CI agree with your decision in step 8? Why or Why not?
Answer:

JMP Instructions:

1. Enter all data into one column, name the column *y*
2. Under *columns*, choose *new column*, name the column *x*
3. Under *data type*, choose *character*
4. Click *okay*
5. Label the data in the "*x*" column – 1's for the first data set, 2's for the second data set
6. Under *Analyze*, choose *fit y by x*
7. Click on *y* column, click on *y*
8. Click on *x* column, click on *x*
9. Choose *normal quantile* plot
10. Choose *unequal variances*
11. Choose *means/anova/pooled t-test*

JMP Output:

Oneway Analysis of final exam scores By x

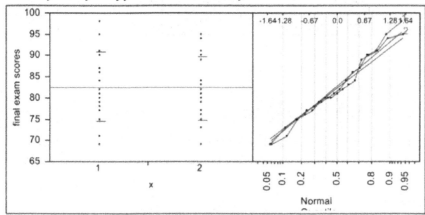

Oneway Anova

t Test

2-1

Assuming equal variances			
Difference	-0.4522	t Ratio	-0.16625
Std Err Dif	2.7201	DF	31
Upper CL Dif	5.0954	Prob > |t|	0.8690
Lower CL Dif	-5.9998	Prob > t	0.5655
Confidence	0.95	Prob < t	0.4345

Hypothesis Testing for Two Parameters   177

Tests that the Variances are Equal

Level	Count	Std Dev	MeanAbsDif to Mean	MeanAbsDif to Median
1	16	8.170832	6.523438	6.437500
2	17	7.454272	5.923875	5.882353

Test	F Ratio	DFNum	DFDen	p-Value
O'Brien[.5]	0.1934	1	31	0.6632
Brown-Forsythe	0.1143	1	31	0.7371
Levene	0.1498	1	31	0.7013
Bartlett	0.1264	1	.	0.7222
F Test 2-sided	1.2015	15	16	0.7181

Checking your Understanding: (Answers are found in the Appendix)

1. Suppose you are interested in determining if men make more money than women, on average. To test, a random sample of ten men and ten women is obtained. At the 5% significance level, can you conclude that the average salary for men is higher than it is for women? The data given below represents the salaries in thousands.

Men	67.441	53.159	56.017	61.264	64.281	61.488	63.665	66.792	63.107	53.636
Women	54.651	61.213	58.26	55.28	57.421	55.102	44.188	49.53	45.767	60.564

Section 3: Homework

Show all work/steps! You must use JMP to analyze the data.

1. Do smokers have higher blood pressure? A random sample of twenty-eight smokers was obtained and their systolic blood pressure was obtained. Another random sample of thirty non-smokers was obtained and their systolic blood pressure was obtained.

Smokers Blood Pressure			Nonsmokers Blood Pressure		
127	134	143	106	116	130
141	135	138	118	121	126
129	137	130	123	118	117
136	135	129	117	118	125
144	138	131	122	112	115
128	136	140	117	111	127
131	133	140	118	128	116
137	133	133	124	119	116
139	131	139	126	116	114
136			125	109	113

    a. At the 5% significance level, what conclusion can be reached?

    b. Is there a possible confounding factor that was not controlled for in the experiment?

178 Introduction to Statistics

2. The BMI levels for a random sample of thirty females with a college degree and for a random sample of thirty-two females with a high school diploma is listed in the table below. At the 5% significance level, can you conclude that the average BMI level for females with a college degree is different than the average BMI level for females with just a high school degree?

BMI Level For H.S. Graduates				BMI Level For College Graduates			
19.1	21.3	21	17.3	21.6	19.4	21.2	23.2
19.8	20	19.2	17.8	22.6	18.8	20.6	21.1
19.8	18.3	18.1	18.8	20.4	20.8	21.2	18.4
19.1	19.2	20.6	17.4	21.4	21.5	22.6	20.1
19.5	17.3	19.5	17.2	21.2	21.3	20.8	21.2
19.6	18.6	18.5	18.5	21.6	20.7	20.5	22.2
17.9	19.3	18.8	18.4	21.4	22.6	21.4	
19.6	19.5	17.4	21.4	19.9	21.7	19.5	

3. Many children and even adults will not eat the bread crust. But is that the best part to eat in terms of the fiber content? It is well known that wheat bread has more fiber than white bread, but what amount of the fiber is contained in the crust? A study was done to determine if eating white bread with the crust is just as good or better for you in terms of fiber content as eating wheat bread without the crust. At the 2.5% significance level, can you conclude that there is a more fiber in the white bread with the crust than there is in wheat bread without the crust?

White With Crust	1.5	1.08	1.11	1.21	1.13	1.3	1.14	1.1	1.11	1.2	
	1.15	1.2	0.97	0.98	1.27	0.88	1.11	0.93	0.88	1.15	
Wheat Without Crust	1.13	1.16	0.73	1.03	0.81	1.17	0.67	1.19	1.15	1.4	1.04
	1.13	1.02	1.33	0.55	1.02	1.47	0.93	1.09	0.94	0.95	0.98

➡ Section 4: Hypothesis Testing for Two Population Proportions for Independent Samples:

When testing two population proportions, our first question should be what it is for testing two population means: "Are the samples independent or dependent?" Because of time, we will only consider the case where the samples are *independent*. As always, we want to choose the appropriate test; the test that controls for alpha and maximizes power. Recall, once again the main idea for hypothesis testing and CI remain the same. We just need to know the following:

- The parameter of interest = _____

- The point estimate = _____

- General form for Null and Alternative hypothesis:

  Answer:

Hypothesis Testing for Two Parameters **179**

- Test Statistic and its distribution:

$$Z_0 = \frac{(\hat{p}_1 - \hat{p}_2) - (p_1 - p_2)}{\sqrt{\bar{p} \cdot \bar{q} \left(\frac{1}{n_1} + \frac{1}{n_2} \right)}} \sim N(0,1)$$

$$\text{where } \bar{p} = \frac{n_1 \hat{p}_1 + n_2 \hat{p}_2}{n_1 + n_2}.$$

- General form of the CI:

$$(\hat{p}_1 - \hat{p}_2) \pm z_{a/2} \sqrt{\frac{\hat{p}_1 \cdot \hat{q}_1}{n_1} + \frac{\hat{p}_2 \hat{q}_2}{n_2}}$$

- Assumptions:

   $n_1{}^*p_1$ and $n_1{}^*q_1$ are both greater than 5 (for both hypothesis test and for CI, you have to use $\hat{p}$ and $\hat{q}$.) $n_2{}^*p_2$ and $n_2{}^*q_2$ are both greater than 5

- JMP does not use this test to analyze two proportions from independent samples. The test it uses is the WALD test, which for two proportions has an approximate normal distribution. When we use JMP, JMP will only give us a p-value and a CI. It does not give us a traditional observed value.

The p-value represents the probability of the populations having the characteristic of interest:

Adjusted Wald Test
P(BIPOLARINOT VCFS)-P(BIPOLARIVCFS) ≥ 0
P(BIPOLARINOT VCFS)-P(BIPOLARIVCFS) ≤ 0
P(BIPOLARINOT VCFS)-P(BIPOLARIVCFS) $= 0$

You choose the p-value that corresponds to the alternative hypothesis.

The CI is a CI on the difference of the two proportions. You have to be careful with JMP because JMP is going to put the characteristics in alphabetical order. So you want to look at the output to see if JMP calculated the CI on $p_1 - p_2$ or on $p_2 - p_1$.

Two Sample Test for Proportions

Description	Proportion Difference	Lower 95%	Upper 95%
P(BIPOLARINOT VCFS)-P(BIPOLARIVCFS)	-0.09929	-0.15663	-0.0411

- Example: Velo-Cardio-Facial Syndrome (VCFS) is a syndrome characterized by having more than 180 clinical features both physical and behavioral including congenital heart disease, immune disorders, feeding problems, cleft palate, development disorders, and psychological disorders such as ADHD, bi-polar disorder, and schizophrenia. Researchers would like to know if the proportion of those with VCFS having bi-polar disorder is greater than the proportion of those without VCFS having bi-polar disorder. A random sample of 400 adults with VCFS was obtained and 114 of them were found to be bi-polar. A random sample of 420 adults without VCFS was obtained and seventy-eight of them were found to be bi-polar. At the 5% significance level, what conclusion can the researchers make?

1.

2.   3.

4.

5. p-value:

6. CI:

7. Decision: (Using p-value and CI)

8. Conclusion:

Hypothesis Testing for Two Parameters **181**

JMP Instructions:

1. Create three columns:

   A *column* describing the two populations. (If you use numbers instead of letters you will have to make the column a "character" column. Ex: VCFS, not VCFS.)

   A *column* describing the characteristic of interest. (If you use numbers instead of letters you will have to make the column a "character" column. Ex: Bipolar, not Bipolar.)

   A *frequency* column. Enter the frequency data. Make sure your numbers match the pop/char of interest. (This column should NOT be a character column.)

2. Under *distributions*, choose fit *y* by *x*.
   Click on the "*population*" column, click on *x*.
   Click on the "*characteristic of interest*" column, click on *y*.
   Click on the "*frequency*" column, click on Freq.

3. Under the red options button, choose *two sample test for proportions*.

- JMP Output:

Contingency Analysis of BIPOLAR By VCFS

Freq: FREQUENCY

Two Sample Test for Proportions

Description	Proportion Difference	Lower 95%	Upper 95%
P(BIPOLARINOT VCFS)-P(BIPOLARIVCFS)	-0.09929	-0.15663	-0.0411

Adjusted Wald Test	Prob
P(BIPOLARINOT VCFS)-P(BIPOLARIVCFS) ≥ 0	0.9996
P(BIPOLARINOT VCFS)-P(BIPOLARIVCFS) ≤ 0	0.0004*
P(BIPOLARINOT VCFS)-P(BIPOLARIVCFS) = 0	0.0008*

Checking your Understanding: (Answers are found in the Appendix)

1. Health researchers are interested in determining if more men smoke than women do. A random sample of 380 men included 97 who smoke, and a random sample of 345 women included 69 who smoke. Use a 0.1 significance level to test the claim that the proportion of men who smoke is greater than the proportion of women who smoke.

Section 4: Homework

Show all work/steps! You must use JMP to analyze the data.

1. Feverfew is an herb that many claim will cure a migraine. Does it work better than standard migraine medicine? To answer this question, one hundred randomly selected migraine sufferers took feverfew at the onset of a migraine. Of the one hundred subjects, eighty-eight reported relief from the migraine. Another random sample of 125 migraine sufferers took the standard migraine medicine at the onset of a migraine. Of those 125 subjects, seventy-six reported relief from the migraine. At the 5% significance level, is there a difference in the proportion of subjects that reported relief from the migraine for the two types of cures: feverfew and standard migraine medicine?

182   Introduction to Statistics

2. Many claim that white noise such as a vacuum cleaner or large box fan will help soothe colicky babies. To test this claim, 130 randomly selected moms of colicky babies rocked their baby in a room with a large box fan turned on high. Of the 130 moms, ninety reported that the large box fan did indeed soothe their baby. Another random sample of 150 moms of colicky babies rocked their baby in a room without any white noise. Of the 150 moms, ninety-four reported that they were able to soothe their baby without using any white noise, simply by rocking them. At the 10% significance level, can you conclude that white noise does help soothe a colicky baby?

3. Many people think of fishing as a boy's activity. But is that true? Do boys like fishing more than girls? To answer this question, a random sample of 200 boys between the ages of eight and twelve was obtained. Of the 200 boys, 124 said they like to fish. Another random sample of 190 girls between the ages of eight and twelve was obtained. Of the 190 girls, fifty-four said they like to fish. At the 2.5% significance level, what conclusion can be reached?

▶ Section 5: Hypothesis Testing for Two Parameters Overview:

Fill in the following for testing two population means from dependent samples:

When performing a hypothesis test for two population means from dependent samples, we analyze the

_____ data. Thus we have a new random variable:

Parameter:

Point Estimate:

General form for Ho:

   Ha:

Test Statistic and its distribution:

Assumptions:

General form of CI:

Fill in the following for testing two population means from independent samples:

Parameter:

Point Estimate:

General form for Ho:

Ha:

Test Statistic and its distribution:

Assumptions:

General form of CI:

General form of the RR (for both methods):

General form of the probability statement of the p-value (for both methods):

Fill in the following for testing two population proportions from independent samples:

Parameter:

Point Estimate:

General form for Ho:

Ha:

Test Statistic and its distribution:

Assumptions:

The _____ Test just gives the p-value and the CI for the difference in the proportions. So,

If CI on $(p_1 - p_2) = (-a, -b)$, what is the decision of the test?

What does that imply about the relationship between p_1 and p_2?

If CI on $(p_1 - p_2) = (a, b)$, what is the decision of the test?

What does that imply about the relationship between p_1 and p_2?

If CI on $(p_1 - p_2) = (-a, b)$, what is the decision of the test?

What does that imply about the relationship between p_1 and p_2?

Chapter 7 Review Homework:

1. There is a theory that people who drink caffeine can run longer than those who do not. To test this theory a random sample of seven people who drink caffeine and a random sample of seven people who do not drink caffeine was obtained and their length of time they could run non-stop was measured, in seconds.

Non-Caffeine drinkers	359	280	138	227	203	184	231
Caffeine drinkers	394	477	439	428	391	488	454

   a. At the 5% significance level, can you conclude that those who drink caffeine can run longer than those who do not?

   b. Are the samples independent or dependent?

   c. If independent, would using dependent samples have been better? Why or Why not? If dependent, explain why dependent samples were used.

2. Are females more likely than males to stay in an unsatisfactory job? To answer this question, Foxworthy News randomly sampled 150 females and 170 males. Sixty females said they have stayed in an unsatisfactory job and sixty-five males said they have stayed in an unsatisfactory job. At the 2.5% significance level, what conclusion can be reached?

3. Do wearing custom-made shoes, made specifically for the runner, help a runner run faster? To answer this question, twelve competitive runners were randomly selected and their maximum velocity (in mph) was recorded while wearing the custom-made shoes and then again while not wearing the custom-made shoe. (A time for rest was granted and it was randomly determined if they ran with the custom-made shoes first or second. The runners ran 100 yards.)

Custom shoes	23.2	20	17.6	22.9	21.5	23	20.8	18.9	24	18.8	19.2	19.3
Regular shoes	14.8	16.1	15.4	18	20.4	18.2	19.3	16.3	19	19.6	18.4	18.5

   a. At the 5% significance level, what conclusion can be reached?

   b. Are the samples independent or dependent?

   c. If independent, would using dependent samples have been better? Why or Why not? If dependent, explain why dependent samples were used.

4. Is there a difference in overall happiness for cat owners versus dog owners. This was a question of interest for psychologist. To study, a random sample of 12 cat owners and a random sample of 12 dog owners was obtained and their score on the Happiness Scale Test was calculated. At the 5% significance level, what conclusion can be reached?

Cat Owners	29.3	21.4	19.4	22.8	17.3	28.2	18.9	25.6	26.2	13.8	21.6	24.4
Dog Owners	18.2	16.8	16.6	17.3	14.5	13.9	11.6	11.8	16.4	15.7	13.4	11.8

Hypothesis Testing for Two Parameters **187**

CATEGORICAL ANALYSIS

Section 1: The Chi-Square Goodness of Fit Test:

There are several situations we have already discussed in which the data we are analyzing is count data, such as when testing one or two population proportions. An area of statistics that deals with count data is called *Categorical Analysis*. There are several tests that deal with count data; one such test is the *Chi-Square Goodness of Fit test*.

There are two uses for the Chi-Square Goodness of Fit Test:

1. When testing a distributional assumption about count data. For example, if you have some data and you want to determine if the data came from a population that has a Binomial or a Poisson distribution.
2. When testing more than two population proportions.

We will consider the second use of the Chi-Square Goodness of Fit Test.

Section 2: Testing More Than Two Population Proportions (The Goodness of Fit Test):

Previously, when we considered other hypothesis testing scenarios, we began with defining the parameter being tested and the point estimate used for testing the parameter. The Chi-Square Goodness of Fit test is not testing an actual parameter; it takes into account what we call the observed and the expected data values. To work the problem, we will follow the same logic as always of hypothesis testing. Thus, in order to know how to work a Goodness of Fit test, we just need to define the observed and expected values, the null and alternative hypothesis, the assumptions, and the test statistic and its distribution.

- The *observed values* are the values actually observed, i.e., our data. We summarize our data (the observed values) into a *one-way table*. If we are testing k population proportions, then there are k cells in the one-way table.

- The *expected values*, denoted as E_i, is what we would expect to see if the null hypothesis is true.

  Thus, $E_i = $ _____ . where $i = 1, ..., k$

- The *null and alternative hypothesis*: we can either assume all the proportions are equal or we can assume that the proportions are equal to some value.

For example:

$H_o: p_1 = p_2 = p_3 = $ _____

H_a: at least one differs from assumed

or

$H_o: p_1 = .25, p_2 = .55, p_3 = .20$

H_a: at least one is different than assumed

- The *test statistic*: The test statistic is the *Chi-Square (χ^2) Goodness of Fit Test*, which has the following formula:

$$\chi^2 = \sum \frac{(O_i - E_i)^2}{E_i} \sim \chi^2_{(k-1)}$$

- The *Chi-Square distribution*:

  1. There are infinitely many different χ^2 distributions; each one is determined by its degrees of freedom. The degrees of freedom vary, depending on the type of problem being tested.
  2. The shape of the distribution depends on the degrees of freedom.

Chi-Square Distributions

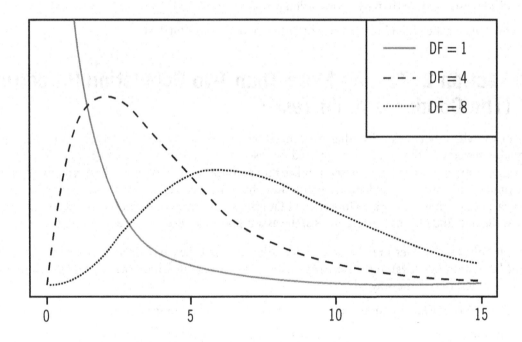

- The assumptions: 1. Have a multinomial experiment
  2. $E_i \geq 5$, for each cell

- The *type of test*: Throughout the test, we always assume that the null hypothesis is true.

  If Ho is true, what does that imply about O_i and E_i?

  Answer:

  What does that imply about the χ^2 value?

  Answer:

  If Ho is false, what does that imply about O_i and E_i?

  Answer:

  What does that imply about the χ^2 value?

  Answer:

  Therefore, is the χ^2 Goodness of Fit test a right-tailed test, a left-tailed test, or a two-tailed test?

  Answer:

  Hence, the rejection region will have the form:

  Answer:

  $x^2(k-1)$

Categorical Analysis 191

And the probability statement for the p-value will always be of the form:

Answer:

- The *Confidence Interval*: The Chi-Square Goodness of Fit test is an overall test. Thus if we reject Ho, that does not imply that all the proportions are different than assumed. It simply implies that at least one proportion is different than assumed.

If we reject Ho, then we must use *Bonferroni Confidence Intervals* to determine which proportion(s) differ from assumed. To construct a Bonferroni CI, we first define a new alpha,

$$\alpha^* = \alpha/k.$$

We then construct a $(1 - \alpha^*)$% CI for *each* proportion. We use $(1 - \alpha^*)$% CI to determine if the proportion is different than assumed. (If there are k proportions being tested, then we will construct k CIs. In reality, we will let JMP do the work for us.)

For us to conclude using a CI that the proportion is different than assumed, then the assumed value of the proportion must be or must not be contained within the CI?

Answer:

Also, although we are having JMP construct $(1 - \alpha^*)$% CIs, we will interpret them as though they are $(1 - \alpha)$% CIs.

- Example: Suppose three college students, Sam, Ann, and Tom, are running for Student Body President. Prior to the election, a survey is conducted to determine the voting preferences of a random sample of 300 college students. At the 5% significance level determine if there is a difference in the voter's preference for the three candidates.

	Candidates	
Sam	Ann	Tom
131	94	75

192   Introduction to Statistics

JMP Instructions:

1. Input the frequencies into Column 1

2. Choose *new column*, under *data type*, choose *character*. Choose *okay*. Label the categories that correspond to the frequencies.

3. Under *Analyze*, choose *Distribution*.

4. Click on the column containing the categories and click on Y

5. Click on the column containing the frequencies and click on *Freq*.

6. Click *OK*.

7. Under the red options button, select *Test Probabilities*.

   A "Test Probabilities" section will be added to the output window.

8. Enter the null probabilities in the cells beneath the heading "Hypoth Prob." Note the categories are in alphabetical order.

9. Click *Done*.

10. Use Pearson statistic and p-value.

11. If necessary, choose the CI, make sure and correctly choose your *Bonferroni* level of confidence.

Categorical Analysis **193**

JMP Output:

Distributions

candidate

Frequencies

Level	Count	Prob
Ann	94	0.31333
Sam	131	0.43667
Tom	75	0.25000
Total	300	1.00000

N Missing

0

3 Levels

Test Probabilities

Level	Estim Prob	Hypoth Prob
Ann	0.31333	0.33333
Sam	0.43667	0.33333
Tom	0.25000	0.33333

Test	ChiSquare	DF	Prob>Chisq
Likelihood Ratio	15.9622	2	0.0003*
Pearson	16.2200	2	0.0003*

Method:

Fix hypothesized values, rescale omitted

Note: Hypothesized probabilities did not sum to 1. Probabilities have been rescaled.

Confidence Intervals

Level	Count	Prob	Lower CI	Upper CI	1-Alpha
Ann	94	0.31333	0.253245	0.380415	0.983
Sam	131	0.43667	0.369957	0.505749	0.983
Tom	75	0.25000	0.195231	0.314136	0.983
Total	300				

Note: Computed using score confidence intervals.

Checking your Understanding: (Answers are found in the Appendix)

1. Is the student grade classification at Baylor the same as it is for other universities? In the United States, of those attending a university, the percentage of freshmen students is 25%, the percentage

of sophomore students is 16%, the percentage of college juniors is 18%, the percentage of seniors is 23%, and the percentage of graduate or "other" students is 18%. For the 2012/2013 school year, Baylor had 3739 freshmen, 2743 sophomores, 2849 juniors, 3399 seniors, and 2634 graduate or "other" students. At the 1% significance level, can you conclude that the proportions for Baylor differ for the US as a whole?

Section 2 Homework:

Show all work/steps! You must use JMP to analyze the data.

1. A researcher is interested in determining if the proportion of children that collect specific types of toys is equal. To test, he takes a random sample of 500 children and finds that 36% of them collect Legos, 19% of them collect Beenie Babies, 18% of them collect Webkinz, and 27% of them collect Hot Wheels. At the 10% significance level, what conclusion can be reached?

2. How do the graduation rates for Baylor compare to other US universities? Nationally, 40% of incoming freshman will receive their undergraduate degree within four years, while 13% of incoming freshman will take five years to graduate and 4% will take six years to graduate. Sadly, 43% of all incoming freshman will not graduate within six years time. From a random sample of 20,000 Baylor students who enrolled as freshmen, 10,800 received their undergraduate degree within four years, 3,200 of them took five years to receive their undergraduate degree, and 400 of them took six years to receive their undergraduate degree. Of the 20,000 surveyed, 5,600 of them did not graduate within six years. At the 5% significance level, what conclusion can be reached?

3. Researchers are interested in determining if there is a difference in the proportion of women who choose the different types of treatments for those diagnosed with Ductal Carcinoma in Situ (DCIS) breast cancer, which is a form of breast cancer that occurs in the Breast Ducts and has not spread. Following a diagnoses of DCIS, women can choose between three different treatment plans: a mastectomy or a lumpectomy followed by radiation therapy or a lumpectomy followed by Adjuvant Endocrine therapy. A random sample of 500 women diagnosed with DCIS breast cancer was obtained. Of the 500 women, 190 chose to have a mastectomy, 150 chose to have a lumpectomy followed by radiation, and 160 chose to have a lumpectomy followed by Adjuvant Endocrine therapy. At the 1% significance level, what conclusion can be reached?

➡ Section 3: Test of Independence:

One of the most frequent uses of the Chi-Square test is for testing whether two criteria of classifications, when applied to a population, are independent.

For example, a market research firm might wish to know whether it can conclude that, for adults in a certain city, the type of beauty care product purchased is associated with the area of residence. This information could help them decide which products to offer and advertise at which store throughout the city.

As before, to work the problem, we just need to know how to arrange the data, the null and alternative hypothesis, the assumptions, and the test statistic and its distribution.

- The *data*: The observed values, i.e., the *data*, is arranged in what is called a *contingency table* or a *two-way table*. The levels of one criterion provide row headings and the levels of the second criterion provide the column headings. The intersection of the rows and columns are called *cells*.

Categorical Analysis **195**

Characteristic 1			
Characteristic 2	**D**	**E**	**F**
A	25	65	42
B	35	22	45
C	62	38	34

- The *hypothesis*:   Ho: independent

    Ha: dependent

- The *test statistic*: The test statistic is the *Chi-Square Test of Independence*, which has the following formula:

$$\chi^2 = \sum \frac{(O_i - E_i)^2}{E_i} \sim \chi^2_{((r-1)(c-1))}$$

where r = # of rows containing data and c = # of columns containing data.

- Finding E_i: Let P_{AD} represent the proportion of people having characteristic A and characteristic D. If the two characteristics are independent, then we know by the multiplication rule for independent events that

$$P_{AD} = P_A \bullet P_D.$$

Thus, since we assume the null hypothesis is true, we know that

$$E_{AD} = P_A \bullet P_D \bullet n$$

Hence, we have the following formula for Ei, for each cell:

$$E_i = \frac{(row\ total)(column\ total)}{grand\ total}$$

- The *assumptions*: $E_i \geq 5$, for each cell

- The *type of test*: Throughout the test, we always assume that the null hypothesis is true. By observing the formula for the test statistic, is the χ^2 *Test of Independence* also a right-tailed test, or is it now a left-tailed test, or a two-tailed test? Why?

Answer:

- Example 1: A market research firm wishes to know whether they can conclude that, for adults in a large metropolitan city, the type of beauty care product purchased is associated with the area of residence. A random sample of 1350 adults from the large metropolitan city was obtained and it was found that of those that live in the city, 143 purchase organic beauty care products, 200

196   Introduction to Statistics

purchase name brand but non-organic beauty care products, and 356 purchase store brand beauty care products. Of the adults that live in the suburbs, 256 purchase organic beauty care products, 175 purchase name brand but non-organic beauty care products, and 220 purchase store brand beauty care products. At the 5% significance level, what conclusion can be reached?

	Type of Residence	
Type of Beauty Product	In the City	In the Suburbs
Organic	143	256
Name Brand	200	175
Store Brand	356	220

JMP Instructions:

1. Input the levels of one of the characteristics, call it *x*. (Make the column a *character* column.)

2. Input the levels of the other characteristic, call it *y*. (Make the column a *character* column.)

3. Input the counts in the third column. Call it *frequency*.

4. Choose *fit y by x*.

5. Select your *x* and *y* and select *freq* for the data.

6. Hit okay.

7. Under red options triangle by the contingency table, choose *expected*.

- JMP Output:

Contingency Analysis of Area of Residence by Beauty Product

Freq: frequency

Contingency Table

Beauty Product by Area of Residence

Count Expected	City	Suburb	Total
Name Brand	200 194.167	175 180.833	375
Organic	143 206.593	256 192.407	399
Store Brand	356 298.24	220 277.76	576
Total	699	651	1350

Tests

N	DF	-LogLike	RSquare (U)
1350	2	32.410930	0.0347

Tests	ChiSquare	Prob>ChiSq
Likelihood Ratio	64.822	<.0001*
Pearson	64.155	<.0001*

- Comments about the Chi-Square Test of Independence:

1. For the Chi-Square Test of Independence, the value of the χ^2 test statistic depends on the way the categories are defined for the two variables. If the area of residence is classified as in the city vs. suburb, then the conclusion of the test might be different than if the categories are classified as north side of city, south side of city, east side of city, and west side of city. A report of the test result should identify the categories for the variables.

2. The Chi-Square Test of Independence treats the classifications as nominal. If the classifications have an understood order (i.e., are ordinal data), the Chi-Square Test of Independence does not take the ordering into account. For example if one characteristic is the "severity" of injury, there is an understood order in the data and one should use a more sophisticated method that takes into account the ordering of the data.

3. The Chi-Square Test of Independence should not be calculated using percentages or rates; it should only be calculated using observed frequencies.

4. The Chi-Square Test of Independence is a large sample test and should not be used for small sample sets. If the sample size assumption is not met, you should use *Fisher's Exact Test.*

5. The Chi-Square test assumes the "levels of the characteristics" are independent. If the columns are dependent samples, such as when each row has the same subjects, then *McNemar's Test* is more appropriate to use.

Checking your Understanding: (Answers are found in the Appendix)

1. A brokerage firm wants to see whether the type of account a customer has between the choices bronze, silver, gold, platinum, or diamond is related to who makes the trade, the customer him/herself or a professional trader hired by the person. A random sample of trades made for its customers over the past year was randomly chosen and it was found that for the trades made by a hired professional broker 55 were at the bronze level, 62 were silver, 75 were gold, 72 were platinum, and 83 were diamond. Of the trades made by the customer him/herself 75 were at the bronze level, 80 were silver, 82 were gold, 65 were platinum, and 67 were diamond. At the 2.5% significance level, what conclusion can be reached?

Section 3 Homework:

Show all work/steps! You must use JMP to analyze the data.

1. Is there a relationship between buying organic produce and having a college degree for males and females.? A random sample of 1,077 U.S. adults was obtained. The contingency table below gives the results of the sample. At the 2.5% significance level, what conclusion can be reached?

	Buys Organic	*Buys Non-Organic*
Females with Degree	125	154
Females without Degree	114	212
Males with Degree	165	152
Males without Degree	85	70

2. For migraine sufferers, is there an association between gender and using a cold or hot compress to treat migraines? From a random sample of 600 migraine sufferers, 192 females received relief using a hot compress while 208 females received relief using a cold compress. Whereas ninety-eight males received relief using a hot compress and 102 males received relief using a cold compress. At the 1% significance level, what conclusion can be reached?

3. Researchers would like to know, for women diagnosed with breast cancer in only one breast, is there a relationship between type of breast cancer a women is diagnosed with and whether they choose to have a single or double mastectomy. To answer, a random sample of 610 women diagnosed with

Categorical Analysis **199**

breast cancer in only one breast was obtained. Of those diagnosed with Intraductal Carcinoma, 70 chose to have a single mastectomy while 45 chose to have a double mastectomy. Of the women diagnosed with Invasive Ductal Carcinoma, 120 chose to have a single mastectomy while 90 chose to have a double mastectomy. For the women diagnosed with Invasive Lobular Carcinoma, 75 opted to have a single mastectomy while 85 had a double mastectomy performed. For women diagnosed with Triple-Negative breast cancer, 40 chose to have a single mastectomy while 85 decided to have a double mastectomy. At the 5% significance level, what conclusion can be reached?

➡ Section 4: Tests of Homogeneity:

In the tests of homogeneity we are testing if several populations are homogeneous or similar with respect to one characteristic.

For example, the local cable provider might want to know whether different age groups are similar (or different) with respect to the type of television programs they prefer to watch.

Once again, to work the problem, we just need to know how to arrange the data, the null and alternative hypothesis, the assumptions, and the test statistic and its distribution.

- The *data*: The observed values, i.e., the *data*, are again arranged in what is called a *contingency table* or a *two-way table*. The levels of the population provide row headings and the levels of the characteristic provide the column headings. The intersection of the rows and columns are called cells.

- The *hypothesis*:   Ho: homogeneous

    Ha: heterogeneous

- The *test statistic*: The test statistic is again the Chi-Square test, which has the same formula:

$$\chi^2 = \sum \frac{(O_i - E_i)^2}{E_i} \sim \chi^2_{((r-1)(c-1))}$$

- Finding E_i: Let E_{AI} represent the number of people in age group I that prefer program A.

    Now P_A represents the proportion of people who prefer program A. If all populations are homogeneous, then we expect this proportion to be true for all three populations. Thus, since there are n_i people in ethnic group I, we have the following formula for E_{AI}:

$$E_{AI} = n_i \bullet P_A$$

    This is the same as before: $E_i = \dfrac{(row\ total)(column\ total)}{grand\ total}$

- The *assumptions*: $E_i \geq 5$, for each cell

- The *type of test*: Throughout the test, we always assume that the null hypothesis is true. By observing the formula for the test statistic, is the χ^2 test still a right-tailed test, or is it now a left-tailed test, or a two-tailed test? Why?

    Answer:

- Example 1: The local cable provider would like to determine whether different age groups are similar with respect to the type of television programs they prefer to watch. A random sample of adults was obtained. Of those that were less than thirty, 200 preferred reality TV, 185 preferred comedies, 150 preferred dramas, and eighty-eight preferred documentaries. Of those that were between thirty and fifty years of age, ninety-two preferred reality TV, 125 preferred comedies, 188 preferred dramas, and 139 preferred documentaries. Of those that were older than fifty, 52 preferred reality TV, one hundred preferred comedies, 220 preferred dramas, and 150 preferred documentaries. At the 2.5% significance level, what conclusion can be reached?

	Type of Show			
Age Group	Reality TV	Comedy	Drama	Documentary
< 30	200	185	150	88
30 – 50	92	125	188	139
> 50	52	100	220	150

JMP Instructions:

1. Input the levels of one of the characteristics, call it *x*. (Make the column a *character* column.)

2. Input the levels of the other characteristic, call it *y*. (Make the column a *character* column.)

3. Input the counts in the third column. Call it *frequency*.

4. Choose *fit y by x*.

5. Select your *x* and *y* and select *freq* for the data.

6. Hit okay.

7. Under red options triangle by the contingency table, choose *expected*.

- JMP Output:

Contingency Analysis of Type of Program by Age

Freq: frequency 2

Contingency Table

Age by Type of Program

Count Expected	Comedy	Documentary	Drama	Reality TV	Total
< 30	185 151.231	88 139.059	150 205.822	200 126.887	623
< 50	100 126.714	150 116.515	220 172.455	52 106.316	522
30 – 50	125 132.054	139 121.426	188 179.723	92 110.797	544
Total	410	377	558	344	1689

Tests

N	DF	-LogLike	RSquare (U)
1689	6	74.450206	0.0322

Tests	ChiSquare	Prob>ChiSq
Likelihood Ratio	148.900	<.0001*
Pearson	146.160	<.0001*

Checking your Understanding: (Answers are found in the Appendix)

Campus life administrators want to determine if students are similar with respect to if and where they work. A random sample of freshman found that 320 do not work, 260 work as a student worker for the University, and 450 work off-campus. A random sample of sophomores found that 280 do not work, 220 work as a student worker for the University, and 350 work off-campus. A random sample of juniors found that 250 do not work, 315 work as a student worker for the University, and 380 work off-campus. A random sample of seniors found that 280 do not work, 220 work as a student worker for the University, and 350 work off-campus. At the 5% significance level what conclusion can be reached?

202 Introduction to Statistics

Section 4 Homework:

Show all work/steps! You must use JMP to analyze the data.

1. Do males and females differ with respect to the type of degree earned? A random sample of 1,005 undergraduates was obtained. Of the males sampled, 200 were earning a Bachelor of Arts (BA) degree while 280 were obtaining a Bachelor of Science (BS) degree. Of the females sampled, 275 were earning a BA degree while 250 were earning a BS degree. At the 10% significance level, what conclusion can be reached?

2. Are ethnic groups similar with respect to the high school sport their daughters prefer to play? A random sample of US adults whose daughters play a sport in high school was obtained. The data below represents the type of sport each daughter plays and their ethnicity. At the 5% significance level, what conclusion can be reached?

	Ethnicity			
Sport	**African American**	**Asian**	**Caucasian**	**Hispanic**
Basketball	175	28	120	85
Soccer	55	43	75	165
Softball	62	48	90	88
Volleyball	78	35	88	65

➡ Section 5: The Chi-Square Test Overview:

Fill in the following for testing more than two population proportion:

Parameter:

E_i:

General form for Ho:

Ha:

Test Statistic and its distribution:

Categorical Analysis **203**

Assumptions:

General form of the RR:

General form of probability statement for the p-value:

Method used to determine which proportions differ:

Construct _____ _____ _____ for each p_i:

Define $\alpha^* =$

If the assumed value of p_i is not contained within the CI, then we _____ .

Fill in the following for testing independence:

Want to Show:

E_i:

General form for Ho:

Ha:

Test Statistic and its distribution:

204   Introduction to Statistics

Assumptions:

General form of the RR:

General form of probability statement for the p-value:

Conclusion if reject Ho:

Conclusion if fail to reject Ho:

Fill in the following for testing homogeneity:

Want to Show:

E_i:

General form for Ho:

Ha:

Test Statistic and its distribution:

Assumptions:

General form of the RR:

General form of probability statement for the p-value:

Conclusion if reject Ho:

Conclusion if fail to reject Ho:

➡️ Chapter 8 Review Homework:

1. Is there a relationship between gender and the type of car driven? A random sample of 1000 adults was obtained. The subjects sampled were asked their gender and what type of car they drive given choices: car, truck, or SUV. Of the females sampled, 210 drive a car, 110 drive a truck, and 180 drive a SUV. Of the males sampled, 175 drive a car, 140 drive a truck, and 185 drive a SUV.

   a. At the 5% significance level what conclusion can be reached?

   b. Does the decision of the test change at the 2.5% significance level?

2. Has the percentage of registered nurses earning more advanced academic degrees changed? In 1980 55% of registered nurses had a high school diploma as their highest level of education, 18% had an associate's degree, 22% had a bachelor's degree, and 5% had a master's degree. In 2012, a random sample of 800 registered nurses was obtained. Of those sampled, 111 had a high school diploma as their highest educational degree, 295 had an associate's degree, 289 had a bachelor's degree, and 105 had a master's degree. At the 5% significance level, what conclusion can be reached?

3. Do newly married couples differ with respect to where they want to live? A random sample of newlyweds was obtained. Of the 500 husbands sampled, 135 wanted to live in the suburbs, 298 wanted to live in the city, and sixty-seven wanted to live in the country. Of the 500 wives sampled, 130 wanted to live in the suburbs, 280 wanted to live in the city, and ninety wanted to live in the country. At the 10% significance level, what conclusion can be reached?

Categorical Analysis **207**

SIMPLE LINEAR REGRESSION

Section 1: Introduction:

In regression, we look to see what type of relationship exists between variables and use the relationship to build a model that can be used for describing, controlling, and predicting.

- A *mathematical* relationship versus a *statistical* relationship:

  In a mathematical relationship, we have an exact relationship between variables: if we know the value of one variable, there is a formula that will give us the value of the other variable.

  In a statistical relationship, we have an approximate relationship between variables: if we know the value of one variable, there is a model that will give us an estimate for the value of the other variable. The stronger the relationship, the better estimates the model will give.

- The type of model built depends upon the type of variables and the type of relationship that exists between the variables.

- Consider the following scatterplots:

Section 2: Basic Properties and Definitions:

- X is called the "independent" or "predictor" variable
- Y is called the "response" or "dependent" variable
- The _____ variable is the variable we want to predict.
- Observed Values: The observed values are the actual known values of Y, given X.

  Notation:

Simple Linear Regression

- Predicted Values: The predicted values, also called the fitted values, are the values of each case that is predicted (or found by) the regression equation.

  Notation:

- Residuals: The residuals are the difference between the observed values and the predicted values.

  Formula:

➡ Section 3: Simple Linear Regression:

- In Simple Linear Regression, there is
  - One Response (dependent) Variable, Y
  - One Predictor (independent) Variable, X
  - The relationship is linear.

The hypothetical Regression Model: $\quad Y = \beta_0 + \beta_1 X + \varepsilon$

where

- β_0 represents the _____ .

- β_1 represents the _____ .

- ε represents the _____ .

- ε is what makes the relationship _____ .

- In regression, for each value of x, there is a subpopulation of y values. We use the notation $\mu_{Y|x}$ to denote the average value of all Y's given $X = x$. On average, the random errors are assumed to be zero.

  - Thus, the average value of Y given $X = x$ is: $\quad \mu_{Y|x} = \beta_0 + \beta_1 x.$

    Hence, the regression line is estimating the *average* value of y given x.

- The graph below depicts what is meant by for each x there is a subpopulation of y values and what we mean by estimating the average value of y:

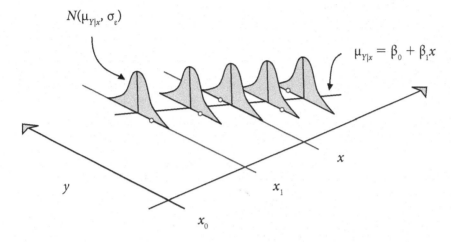

- In practice, we sample for each x, one y. A picture of this is called a scatterplot:

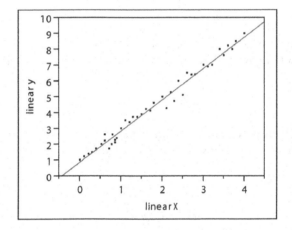

- The sample regression line is the line, $\hat{y} = b_0 + b_1 x$, (also called the line of best fit) where

    b_0 represents _____.

    b_1 represent: _____.

- What is the relationship between the population regression line and the sample regression line? The parameters in the hypothetical model need to be estimated...

$$\mu_{Y|x} = \beta_0 + \beta_1 x$$
$$\hat{y} = b_0 + b_1 x$$

- $\hat{y}$ is a _____ for $\mu_{Y|x}$

- b_0 is a _____ for β_0

- b_1 is a _____ for β_1

Simple Linear Regression

- To find the sample regression line we want to choose the line that best "fits" the data. What exactly is meant by "best fits the line"?

  It means, we choose the line that minimizes the residuals. Recall,

  Residual = observed value − predicted value = $y - \hat{y}$

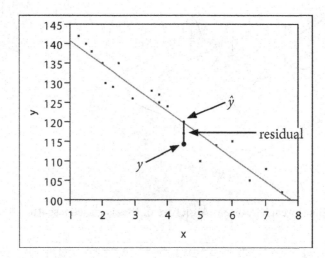

Each point has a residual, so to find the regression line, we want to find the line that minimizes *all* the residuals, which you might think would lead to minimizing the equation:

$$\sum_{i=1}^{n}(y_i - \hat{y}_i)$$

However, this sum will always equal zero. This is because the mean is the true center of the data. However much the values deviate above the mean will always equal with however much they deviate below the mean.

Thus, we have to minimize the equation:

$$\sum_{i=1}^{n}(y_i - \hat{y}_i)^2 = \sum_{i=1}^{n}(y_i - (b_0 + b_1 x_i))^2$$

To minimize the equation, we use differential calculus, i.e., we take the derivative with respect to b_0 and with respect to b_1, set equal to zero and solve. This gives us the following sample regression line.

- $\hat{y} = b_0 + b_1 x$ where $b_0 = \bar{y} - b_1 \bar{x}$ and $b_1 = \dfrac{SS_{xy}}{SS_{xx}}$

  and

  $SS_{xx} = \sum_{i=1}^{n} x_i^2 + \dfrac{1}{n}(\sum_{i=1}^{n} x_i)^2$ and $SS_{xy} = \sum_{i=1}^{n} x_i y_i + \dfrac{1}{n}(\sum_{i=1}^{n} x_i)(\sum_{i=1}^{n} y_i)$.

- We will use JMP to find the regression line, but we need to know how to we interpret the values.
- *What is the interpretation of the slope?*
  Answer:

- *What is the interpretation of the y-intercept?* The *y*-intercept may not always have an interpretation. When it does have an interpretation, it means:
  Answer:

  When will it not have an interpretation?
  Answer:

Extrapolation: When the model is used to estimate *y* for values of *x* that are "outside" the range of values considered when building the model.

Suppose you have the following scatterplot and you want to predict the value of *y* for the given *x*.

What would that estimate be?

What if the relationship changes?

The range of x values considered

Simple Linear Regression 213

- Simple Linear Regression Example:

    Is there a relationship between time one spends in strenuous exercise and blood pressure? Researchers would like a model that will estimate systolic blood pressure given amount of time (in hours) a person spends in strenuous exercise per week. To create the model and hence answer the question, a random sample of twenty adults was obtained. Use the data to determine if the relationship is linear and if so, to build the model.

Time	Pressure	Time	Pressure
1.67	138	2	135
1.5	140	2.3	129
2.92	126	3.5	128
2.08	130	2.5	135
1.25	142	3.75	127
4.5	120	3.75	125
5	110	6	115
7	108	4	124
5.5	114	6.5	105
4.5	117	7.5	102

- Which variable is the independent variable?

    Answer:

- Which variable is the dependent variable?

    Answer:

- Is the relationship linear?

    Answer:

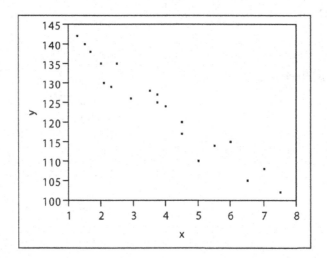

- What is the regression line?

  Answer:

- Interpret the coefficients of the regression line:

  Answer:

  $b_1 =$

  $b_0 =$

  Did we consider values of x near zero?

- Use the model to estimate the systolic blood pressure for someone who exercises an hour and a half five days a week.

  Answer:

- Interpret the estimate found above:

Simple Linear Regression   **215**

> Instructions for using JMP to analyze the data:
>
> 1. Enter the data into two separate columns.
> 2. Click on the dependent variable and click on the *Y, Response* button. Click on the independent variable and click on the *X, Factor* button. Click *okay*.
> 3. Choose *Fit Y by X* under the Analysis window.
> 4. Using the options button, choose *fit line*.

- JMP Output:

Bivariate Fit of y By x

—— Linear Fit

Linear Fit
y = 146.8355 - 6.0050193*x

Summary of Fit

RSquare	0.934533
RSquare Adj	0.930895
Root Mean Square Error	3.102798
Mean of Response	123.5
Observations (or Sum Wgts)	20

Parameter Estimates

Term	Estimate	Std Error	t Ratio	Prob>|t|
Intercept	146.8355	1.612659	91.05	<.0001*
x	-6.005019	0.374622	-16.03	<.0001*

Checking your Understanding: (Answers are found in the Appendix)

A pharmaceutical company has developed a new drug for patients with Wolfe-Parkinson-White syndrome designed to control paroxysmal supraventricular tachycardia (PSVT). Patients with PSVT have a heart rate of more than 100 beats per minute. (BPM) Suppose you are interested in modeling the relationship between the decrease in pulse rate in BPM and dosage in cc. Different doses were administered to 16 randomly selected patients and 20 minutes later the decrease in each patient's pulse rate was recorded. The data collected is given below.

Dosage	2	4	1.5	1	3	3.5	2.5	3	3.5	1.5	2	2	3.75	3.5	1.75	1.25
Drop in HR	14	24	11	6	18	22	17	15	22	9	16	14	23	25	12	12

   a. Which variable is the independent variable?

   b. Which variable is the dependent variable?

   c. Is a linear model appropriate? Explain your answer.

   d. If a linear model is appropriate, find the line of best fit.

   e. If appropriate, interpret the coefficients of the model.

   f. If appropriate, estimate the drop in pulse rate for a dosage level of 2.75 cc. Interpret the estimate.

   g. If appropriate, estimate the drop in pulse rate for a dosage level of 5.5 cc. Interpret the estimate.

Section 3: Homework

1. An insurance salesman would like a model to predict what a healthy person's monthly health insurance premium would be given his or her age. The data listed below gives the monthly premium and age for 28 randomly selected healthy adults.

Age	Premium		Age	Premium
22	154		39	169
23	153		38	170
25	152		40	172
26	153		41	170
29	156		54	221
31	158		56	225
42	182		52	220
43	180		59	232
46	179		57	230
52	205		58	240
57	210		60	249
58	215		48	200
32	161		47	180
35	160		50	191

   a. Which variable is the independent variable?

   b. Which variable is the dependent variable?

   c. Is a linear model appropriate? If so, justify. If not, what type of model, if any, would be most appropriate?

Simple Linear Regression **217**

2. Does gaining weight increase systolic blood pressure for adult females? And, if so, what is the relationship? This was a question of interest for researchers. A random sample of twenty female adults was obtained. The data below is the weight measure in pounds and systolic blood pressure measured in mmHg.

Weight	Blood Pressure		Weight	Blood Pressure
180	138		145	120
175	135		156	125
195	140		135	110
171	129		140	115
160	126		115	108
158	128		149	124
162	130		135	114
150	135		130	105
194	142		130	117
150	127		120	102

a. Which variable is the independent variable?

b. Which variable is the dependent variable?

c. Is a linear model appropriate? If so, justify. If not, what type of model, if any, would be most appropriate?

3. Is a college student's freshman GPA a good predictor for the student's GPA at graduation? To investigate this question, thirty students were randomly selected. The data below represents their GPA at the end of their freshman year and at graduation.

Freshman GPA	Graduation GPA		Freshman GPA	Graduation GPA		Freshman GPA	Graduation GPA
2.32	2.49		3.86	3.38		3.12	3
2.36	3.15		2.97	3.05		1.28	2.64
1.98	3.05		1.88	3.26		1.79	2.9
0.98	2.28		1.67	2.78		1.45	2.7
3.04	3.25		1.54	1.98		3.9	3.85
3.22	2.88		2.88	3.28		4	3.9
2.50	3.08		3.05	2.98		2.88	3.02
2.08	2.28		3.17	2.88		1.42	2.98
3.05	2.38		3.85	3.05		3.8	2.2
3.48	2.68		3.92	3.28		2.5	2.9

a. Which variable is the independent variable?

b. Which variable is the dependent variable?

c. Is a linear model appropriate? If so, justify. If not, what type of model, if any, would be most appropriate?

4. A question of interest is whether increasing the amount of exercise will decrease blood glucose levels for pre-diabetic adults and if so, in what fashion. A random sample of twenty pre-diabetic adults was obtained. The data below gives the amount of time spent exercising per week (in hours) and the person's blood glucose level (measured in mg/dL).

Exercise Time	Glucose Levels		Exercise Time	Glucose Level
1.67	155		4.5	105
2	150		3.75	125
1.5	156		5	95
2.3	149		6	98
2.92	151		7	100
3.5	129		4	118
2.08	145		5.5	114
2.5	135		6.5	98
1.25	154		4.5	107
3.75	127		7.5	80

   a. Which variable is the independent variable?

   b. Which variable is the dependent variable?

   c. Is a linear model appropriate? If so, justify. If not, what type of model, if any, would be most appropriate?

5. Answer the following questions about β_0:

   1. What does the symbol β_0 represent?

   2. Does β_0 always have an interpretation? If not, explain when it does not have an interpretation.

   3. If it does have an interpretation, what is the general interpretation of β_0?

   4. What is the point estimate for β_0?

6. Answer the following questions about β_1:

   a. What does the symbol β_1 represent?

   b. Does β_1 always have an interpretation? If not, explain when it does not have an interpretation.

   c. If it does have an interpretation, what is the general interpretation of β_1?

   d. What is the point estimate for β_1?

7. Answer the following questions about $\mu_{Y|x}$:

   a. What does the symbol $\mu_{Y|x}$ represent?

   b. What is the point estimate for $\mu_{Y|x}$?

8. Consider again the data describing the relationship between weight and blood pressure described in problem 2. Use the data in problem 2 to answer the following questions.

   a. Find the regression line.

   b. Interpret, if possible, the coefficients of the regression line.

   c. If possible, estimate the average systolic blood pressure for an adult who weighs 165 pounds.

   d. If possible, estimate the average systolic blood pressure for an adult who weighs 295 pounds.

Simple Linear Regression **219**

9. Consider again the data describing the relationship between exercise and glucose levels described in problem 4. Use the data in problem 4 to answer the following questions.

   a. Find the regression line.

   b. Interpret, if possible, the coefficients of the regression line.

   c. If possible, estimate the average blood glucose level for an adult who exercises two hours a day, every day of the week.

   d. If possible, estimate the average blood glucose level for an adult who exercises one hour a day, three days a week.

➡ Section 4: Confidence Intervals:

For estimating parameters, the regression line, or "line of best fit," is, in essence, a point estimate for the parameters. i.e.,

$$\mu_{Y|x} = \beta_0 + \beta_1 x$$

$$\hat{y} = b_0 + b_1 x$$

Instead of just using the point estimates, we might want more of an idea as to what we can expect to see from the data. Using just a point estimate does not take into account the variability in the data. Thus we might want to calculate a confidence interval for the slope, β_1, and the average of Y, given x, $\mu_{Y|x}$.

- *Confidence Interval for the Population Slope in Simple Regression*: To have more of an idea about the average change in Y for a unit change in x, we need a confidence interval for the slope, β_1.

  Recall, the general form of a CI, is as follows:

  $$\text{pt. est.} \pm \text{RF*SE}$$

  Filling in the values, we get:

  We use $n - 2$ degrees of freedom for the t-table and obtain the standard error of the slope, S_{b_1}, from the JMP output.

Parameter Estimates				
Term	Estimate	Std Error	t Ratio	Prob>ltl
Intercept	146.8355	1.612659	91.05	<.0001*
x	-6.005019	0.374622	-16.03	<.0001*

- Example 1: In the blood pressure vs. exercise time example find a 95% CI for the population slope:

  - pt. est.:   - SE:

220   Introduction to Statistics

- RF:

So a 95% CI for β_1 is:

- Interpretation:

We are 95% confident that if we _____ the amount of time spent in strenuous exercise by _____, the _____ systolic blood pressure will _____ between _____ and _____ mmhg, on average.

- Use the CI on β_1 to verify if there is a linear relationship between the variables:

Since the CI does not contain the value _____, we can conclude that there is a linear relationship between hours per week of strenuous exercise and systolic blood pressure levels.

$$\mu_{Y|x} = \beta_0 + \beta_1 x$$

We can also do a hypothesis test to determine if there is a linear relationship between the variables.

Parameter Estimates

Term	Estimate	Std Error	t Ratio	Prob>ltl
Intercept	146.8355	1.612659	91.05	<.0001*
x	-6.005019	0.374622	-16.03	<.0001*

Testing: Ho: $\beta_1 = 0$

Ha: $\beta_1 \neq 0$

Since the p-value is less than α, we _____ . Thus we can conclude that $\beta_1 \neq 0$. Hence we can conclude that there is a linear relationship between the two variables.

Simple Linear Regression **221**

- Confidence Interval for $\mu_{Y|x}$: We want an interval estimate of the average value of Y given a value of x; that is, on $\mu_{Y|x}$. Again, the general form of a CI:

$$\text{pt. est.} \pm RF*SE$$

point estimate: the regresssion line evaluated at the point x.

t-value: t-value with n-2 df

standard error of $\hat{y}$:

$$S\sqrt{\frac{1}{n} + \frac{(x - \bar{x})^2}{(n - 1)*Var(X)}}$$

Thus, the formula for a CI on the average value of Y given x is

$$\hat{y} \pm t_{(n-2)} *S* \sqrt{\frac{1}{n} + \frac{(x - \bar{x})^2}{(n - 1)*Var(X)}}$$

JMP will find the CI for us.

- Example 2: Find a 95% CI on the average systolic blood pressure given the person exercises strenuously for forty-five minutes a day, four days a week.

  What is x?

  Answer:

We input this value of x into our data in JMP:

19	4.5	117
20	7.5	102
21	3	.

Leave the y blank. JMP will put the period there for you.

Under the *linear fit* option, choose _____ .

Output:

222   Introduction to Statistics

19	4.5		117	118.27720055	121.54830996
20	7.5		102	98.60171454	104.994006
21	3		•	127.20459974	130.43629445

Thus, the 95% CI for $\mu_{Y|x}$ is:

Interpretation:

We _____, with 95% confidence, that _____ who works out strenuously for _____ a week will have an _____ systolic blood pressure between _____ and _____ mmhg, on average.

- A Prediction Interval, PI, for the value of y given x (*not* the average of y, the predicted value of y) is found using the formula:

$$\hat{y} \pm t_{\alpha/_2} S \sqrt{1 + \frac{1}{n} + \frac{(x - \bar{x})^2}{(n-1)*Var(X)}}$$

JMP will also find the PI for us.

Example 3: Suppose we want a 95% prediction interval for the systolic blood pressure for a specific person who works out for forty-five minutes a day, four days a week.

JMP Instructions:

Under the *linear fit* option, choose _____

JMP Output:

7.5		102	98.60171454	104.994006	94.537745637	109.0579749
3		•	127.20459974	130.43629445	122.10443045	135.53646374

Interpretation:

We _____, with 95% confidence, that this _____ _____ who works out strenuously for _____ hours a week will have a systolic blood pressure between _____ and _____ mmhg, on average.

Simple Linear Regression **223**

Note that a confidence interval gives the average systolic blood pressure expected for *all* subjects who exercise 45 minutes per day, four days a week, whereas the prediction interval gives the systolic blood pressure expected for a *specific* subject who exercises forty-five minutes a day, four days a week.

The prediction interval is _____ than the confidence interval, this is because one value will vary more (i.e., have more possible values) than several values will.

Checking your Understanding: (Answers are found in the Appendix)

A pharmaceutical company has developed a new drug for patients with Wolfe-Parkinson-White syndrome designed to control paroxysmal supraventricular tachycardia (PSVT). Patients with PSVT have a heart rate of more than 100 beats per minute. (BPM) Suppose you are interested in modeling the relationship between the decrease in pulse rate in BPM and dosage in cc. Different doses were administered to 16 randomly selected patients and 20 minutes later the decrease in each patient's pulse rate was recorded. The data collected is given below.

Dosage	2	4	1.5	1	3	3.5	2.5	3	3.5	1.5	2	2	3.75	3.5	1.75	1.25
Drop in HR	14	24	11	6	18	22	17	15	22	9	16	14	23	25	12	12

  a. Find and interpret a 99% confidence interval for the slope of the regression line.

  b. Find and interpret a 99% confidence interval for the average drop in pulse rate for a dosage level of 2.75 cc.

  c. Find and interpret a 99% prediction interval for a dosage level of 2.75 cc.

Section 4: Homework

  1. For each of the following confidence intervals for β_1, (i) determine whether there is a linear relationship between the two variables and (ii) if there is a linear relationship between the two variables determine whether y is increasing or decreasing as x increases.

  a. (–3.24, –1.23)   b. (–5, 2)   c. (2.86, 5.92)

  2. Once again consider the data describing the relationship between weight and blood pressure described in problem 2. Use the data in problem 2 to answer the following questions.

  a. Find *and* interpret a 95% confidence interval for the population slope.

  b. Find *and* interpret a 95% confidence interval for the average systolic blood pressure for female adults that weigh 158 pounds.

  c. Find *and* interpret a 95% prediction interval for a female adult who weighs 158 pounds.

  d. Explicitly explain the difference between the CI found in part b and the PI found in part c.

  3. Consider once again the data describing the relationship between exercise and glucose levels described in problem 4. Use the data in problem 4 to answer the following questions.

  a. Find *and* interpret a 99% confidence interval for the population slope.

  b. Find *and* interpret a 99% confidence interval for the average blood glucose level for adults that exercise an hour a day, three days a week.

224   Introduction to Statistics

c. Find *and* interpret a 99% prediction interval for an adult that exercises an hour a day, three days a week.

d. Explicitly explain the difference between the CI found in part b and the PI found in part c.

4. For the following two intervals, determine which one is the CI and which one is the PI. Note that all other information was kept the same when constructing the intervals.

Interval 1: (145, 172)   **Interval 2:** (154, 162)

➡ Section 5: Correlation Analysis:

Used for describing the strength of the *linear* relationship between the response and independent variables.

There are two values we can calculate to describe the strength of the linear relationship between two variables.

1. The Correlation Coefficient
2. The Coefficient of Determination

- *The Correlation Coefficient*
- The correlation coefficient measures the strength of the linear relationship between two variables. The correlation coefficient is scaleless and denoted by the symbol, r.

- The formula for finding r: $\quad r = \dfrac{SS_{xy}}{\sqrt{SS_{xy}SS_{xx}}}.$

- The following plots illustrate various values of r. The value of r is written below each plot.

$r = 0.83$

$r = 0.08$

$r = -0.92$

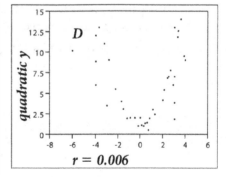
$r = 0.006$

• Which plot represents the strongest linear relationship?

  Answer:

• Which plot represents no relationship?

  Answer:

• Is there a strong relationship being displayed in plot D?

  Answer:

• Is it linear?

  Answer:

• What is the relationship between r and b_1?

  Answer:

- The strength of the linear relationship is represented by how far r is from 0:

  • _____ $\leq r \leq$ _____

  • If $r =$ _____ →

  • If $r =$ _____ → perfect linear relationship, which means graphically:

    Answer:

- The sign of r relates information about the type of linear relationship between the two variables:

  If the value of r is positive → positive or direct linear relationship.
  Thus *high* values of one variable are paired with *high* values of the other

  If the value of r is negative → negative or indirect relationship
  Thus *high* values of one variable are paired with *low* values of the other

- The value of r is easy to interpret when it is close to zero or positive or negative one, but for "middle" values of r, it can be subjective to interpret. In other words, is it a somewhat strong, fairly strong, somewhat weak, fairly weak, etc., linear relationship?

- *The coefficient of determination*:

  The coefficient of determination describes the percentage of the variation in the dependent variable that can be accounted for by the linear relationship between x and y and is denoted by r^2.

  - The formula for finding r^2: $\quad r^2 = \dfrac{SSR}{SS_{yy}} = 1 - \dfrac{SSE}{SS_{yy}}$.

  - In simple linear regression, r^2 is simply the square of r.

226 Introduction to Statistics

- Thus, _____ ≤ r^2 ≤ _____.
- There are two ways to interpret r^2: Suppose $r^2 = 0.9453$.

1. **Variance Interpretation:** Approximately 94.53% of the variation in y can be explained by the linear relationship between x and y.

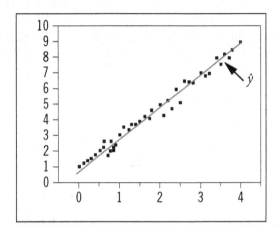

2. **Error Interpretation:** We will reduce our prediction error using the linear regression equation to predict y versus using $\bar{y}$ by 94.53%.

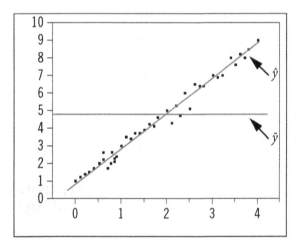

Simple Linear Regression 227

- The following plots illustrate the coefficient of determination. Note that the stronger the linear relationship is between the two variables, the closer the coefficient of determination is to 1. If there is little to no linear relationship, the coefficient of determination is near 0.

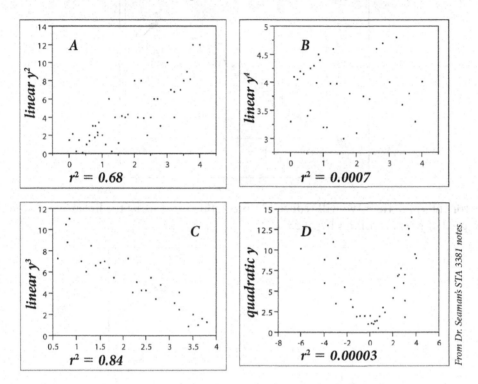

- Example of finding r and r^2
    - For our exercise vs blood pressure example: We get r^2 from JMP. We can calculate r, by taking the square root of r^2. We know that the sign of r has to match the sign of b_1. So, we have $r^2 = 0.934533$

    Variance Interpretation: Approximately 93.45% of the variation in _____ can be explained by the _____ relationship between _____ and _____.

    Error Interpretation: We will _____ the _____ _____ by approximately _____ using _____ to estimate _____ versus using _____ to estimate _____.

    Thus, $r = \sqrt{0.934533} =$ _____.

    (r is _____, since b_1 is _____).

- The correlation coefficient, r, and the coefficient of determination, r^2, describe the strength of the linear relationship, but does the slope, b_1, tell us anything about the strength of the linear relationship?

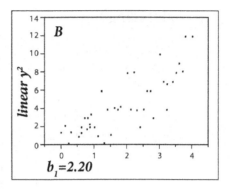

- Which graph represents the variables that have the strongest linear relationship?
  Answer:

- Which regression line has the largest slope?
  Answer:

- So, does the size of the slope have anything to do with the strength of the linear relationship?
  Answer:

- Some comments about correlation:
  - Always, always, always, look at the scatter plot! Why?
    Answer:

  - There is more than one way to measure correlation. The *Pearson product moment correlation coefficient* is the measure we are using and is the most widely used one.
  - Correlation is a dimensionless measure of the strength of the linear relationship. That is, it does not matter what the unit of measurement is for either variable.
  - Correlation measures the strength of the *linear* relationship between variables.
  - **Correlation does not imply causation!!** *Just because two variables have a strong correlation does NOT mean that a cause-effect relationship exists.* If two variables are highly correlated, it *could* mean that the independent variable is causing the dependent variable to occur, but it does not, on its own, imply that the independent variable is causing the dependent variable to occur. There could be another variable that is causing the independent and dependent variable to behave in the same way.

    For example: Ice cream consumption has a positive correlation with drowning rates. Does that mean eating ice cream causes drowning?

Simple Linear Regression

Checking your Understanding: (Answers are found in the Appendix)

A pharmaceutical company has developed a new drug for patients with Wolfe-Parkinson-White syndrome designed to control paroxysmal supraventricular tachycardia (PSVT). Patients with PSVT have a heart rate of more than 100 beats per minute. (BPM) Suppose you are interested in modeling the relationship between the decrease in pulse rate in BPM and dosage in cc. Different doses were administered to 16 randomly selected patients and 20 minutes later the decrease in each patient's pulse rate was recorded. The data collected is given below.

Dosage	2	4	1.5	1	3	3.5	2.5	3	3.5	1.5	2	2	3.75	3.5	1.75	1.25
Drop in HR	14	24	11	6	18	22	17	15	22	9	16	14	23	25	12	12

a. Find the correlation coefficient. Is there a strong linear relationship? Is it direct or indirect?

b. Find the coefficient of determination.

c. Interpret the coefficient of determination using the variance interpretation.

d. Interpret the coefficient of determination using the error interpretation.

Section 5: Homework

1. What symbol is used to represent the correlation coefficient?

2. What symbol is used to represent the coefficient of determination?

3. Suppose you are given the four coefficient of determination values below. Determine which one represents the two variables that are most linearly related.

   Value 1: 0.89   **Value 2:** −0.92   **Value 3:** 0.01   **Value 4:** 0.54

4. Suppose you are given the four correlation coefficient values below. Determine which one represents the two variables that are most linearly related.

   Value 1: 0.89   **Value 2:** −0.92   **Value 3:** 0.01   **Value 4:** 0.54

5. For each of the following correlation coefficient values, determine whether you can conclude that the two variables are linearly related and if so, if the value of y increases or decreases as x increases.

   a. −0.95   b. 0.001   c. 0.92   d. −0.00005

6. Once again consider the data describing the relationship between weight and blood pressure described in problem #2. Use the data in problem #2 to answer the following questions.

   a. Find *and* interpret the correlation coefficient.

   b. Find the coefficient of determination.

   c. Interpret the coefficient of determination using the "variance" interpretation.

   d. Interpret the coefficient of determination using the "error" interpretation.

7. Consider once again the data describing the relationship between exercise and glucose levels described in problem 4. Use the data in problem 4 to answer the following questions.

   a. Find *and* interpret the correlation coefficient.

   b. Find the coefficient of determination.

   c. Interpret the coefficient of determination using the "variance" interpretation.

   d. Interpret the coefficient of determination using the "error" interpretation.

230   Introduction to Statistics

➡ Section 6: Residual Analysis:

We know in regression we are modeling a statistical relationship, not a mathematical relationship. Furthermore, the stronger the linear relationship, the better the model will be in predicting the average value of y. In addition to verifying that there is a linear relationship between the variables, there are other assumptions that must be verified to ensure the validity of the model.

Residual analysis is used to determine whether the linear model is appropriate.

- Recall, the population model for simple linear regression: $y = \beta_0 + \beta_1 x + \varepsilon$

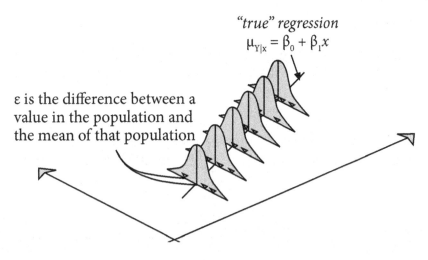

From Dr. Seaman's STA 3381 notes.

- **Simple Linear Regression Assumptions**: The assumptions on the errors, ε:

    1. The relationship between the independent variable, x, and the averages of the dependent variable, y, given x is <u>linear</u>; i.e., $E(Y|x) = \beta_0 + \beta_1 x$.

    2. The errors are independent.

    3. The errors have constant variance (homoscedasticity) and zero mean for all values of x

    4. The errors are normally distributed.

    We can verify assumptions 1 – 3 by creating a residual plot. We can verify assumption 4 by looking at a normal quantile plot of the residuals. Specific properties of the residuals are what allow us to check the assumptions of linear regression by looking at the residuals.

    Recall, each value has a residual which is defined as $res_i = y_i - \hat{y}_i$.

- Properties of residuals

    1. The average of the residuals is always zero. Why?

        Answer:

Simple Linear Regression

2. If assumptions 1 – 3 are valid, then the residuals graphed on a residual plot should be randomly distributed about zero. In other words, there will not be a pattern in the residuals. If there is a pattern, the type of pattern can tell us what assumption is violated. (Technically, the residuals are not independent. However, except for small data sets and residuals for extreme data points, the correlation among residuals is very small and can be ignored.)

3. If assumption 4 on the last page is valid, then the residuals should be closely clustered along the line on the QQ-plot.

- So, to check our assumptions, we need to do the following:

1. Look at the residual plot.

2. Look at the QQ-Plot of the residuals.

3. Check for any outliers.

- Residual plots: Plot the residuals vs. the predicted values and look for patterns. The graphs below illustrate typical patterns.

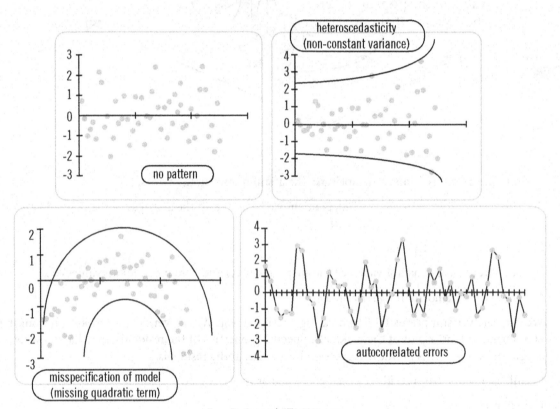

From Dr. Seaman's STA 3381 notes.

- **Step 1: The Residual Plot:**

To get the residual plot, under the linear fit option, choose *plot residual*.

The residual plot for our blood pressure data:

Are any of assumptions 1–3 violated?

Diagnostics Plots
Residual by Predicted Plot

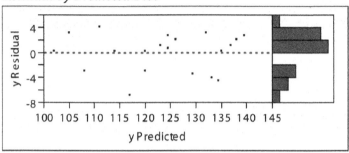

- **Step 2:** The QQ-Plot of the residuals

We look at residuals to see if there is any evidence that the normal error assumption has been violated. We do this with the normal probability plot provided by JMP. Recall, to assume the residuals are normally distributed; the residuals need to be closely clustered to the line.

A Normal Quantile Plot of the residuals is given when the plot residual request is chosen.

Here is the Normal Quantile plot of the residuals:

Can we assume the residuals are normally distributed?

Diagnostics Plots
Residual Normal Quantile Plot

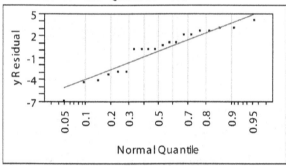

- **Step 3:** Checking for Outliers

If the residuals are normally distributed, then by the Emperical Rule, we know that about 99.7% of the residuals will fall within three standard deviations of the mean.. Thus, since the mean of the residuals equals zero, any residual beyond $\pm 3S = \pm 3(\text{Root MSE})$ is an outlier. Hence, the data value corresponding to that residual is an outlier.

To determine if there are any outliers, we look at the list of residuals provided by JMP. We need to determine if there are any residuals bigger in magnitude than ± 3(Root MSE).

So for our strenuous exercise vs. systolic blood pressure example:

$\pm$ 3(Root MSE) = $\pm$ 3(_____) = _____.

Simple Linear Regression 233

Are there any outliers?

The Residuals:

1.192877243	2.1820625523	0.1870818469	1.1845721996
0.1745336103	−4.345064846	0.6833173759	0.1921011416
2.1720239629	3.1770432576	−6.810408506	−2.802879564
−4.023960601	2.6707691393	4.1946107889	−2.812918153
−3.300848639	2.6833173759	3.1996300836	0.2021397309

Outliers should always be investigated and never simply deleted! If an outlier is not an error, then that outlier might be trying to provide us with information about some phenomenon occurring. If you do omit the outliers and refit the model after removing the outliers, you should always report the fact that you have done so.

- Some other variables in regression to be aware of:
  - Influential Observations: An influential observation is an observation that, although not an outlier, does influence the regression line, which in turn influences the predictions found.
  - How would the observation, x, influence the following regression line?

  Answer:

An influential observation can also have an effect on the correlation coefficient. What is that effect?

Answer:

- Lurking Variables: A lurking variable is a variable that you have not considered in your analysis but which does influence the dependent variable in a regression. Lurking variables also affect correlation. A lurking variable has the same effect on the regression model as an extraneous variable has on an experiment.

  Consider the following example courtesy of Dr. John Seaman:

  Suppose you work for a retail concern and you have studied the relationship between your advertising expenditures and sales. You found it to be strong and positive and conclude that the recent increases in advertising outlays have paid off. In reality, your only significant competitor has been hampered by nearby road construction restricting customer access to his establishment. "Ease of customer access" was a lurking variable in this case.

Checking your Understanding: (Answers are found in the Appendix)

A pharmaceutical company has developed a new drug for patients with Wolfe-Parkinson-White syndrome designed to control paroxysmal supraventricular tachycardia (PSVT). Patients with PSVT have a heart rate of more than 100 beats per minute. (BPM) Suppose you are interested in modeling the relationship between the decrease in pulse rate in BPM and dosage in cc. Different doses were administered to 16 randomly selected patients and 20 minutes later the decrease in each patient's pulse rate was recorded. The data collected is given below.

Dosage	2	4	1.5	1	3	3.5	2.5	3	3.5	1.5	2	2	3.75	3.5	1.75	1.25
Drop in HR	14	24	11	6	18	22	17	15	22	9	16	14	23	25	12	12

Perform residual analysis to verify the assumptions of the model.

Section 6: Homework

1. State the assumptions of the linear regression model.

2. Once again consider the data describing the relationship between weight and blood pressure described in problem 2. Use the data in problem 2 to answer the following questions.

   a. Perform the residual analysis to verify the assumptions of the model. If any assumptions are violated, clearly state which assumption(s) are violated.

   b. Determine if there are any outliers. If there are outliers, state which data value is an outlier.

3. Consider once again the data describing the relationship between exercise and glucose levels described in problem 4. Use the data in problem 4 to answer the following questions.

   a. Perform the residual analysis to verify the assumptions of the model. If any assumptions are violated, clearly state which assumption(s) are violated.

   b. Determine if there are any outliers. If there are outliers, state which data value is an outlier.

4. Define the following terms:

   a. A lurking variable.

   b. An influential observation.

➡ Section 7: Linear Regression and Correlation Analysis Overview:

In Simple Linear Regression, the goal is to build a model that represents the relationship inherent between the two variables. The model can then be used for describing the relationship between the two variables and predicting.

Simple Linear Regression **235**

The X variable is called the _____ or _____ variable.

The Y variable is called the _____ or _____ variable.

The variable we desire to predict or estimate is the _____ variable.

The line of best fit (or the sample regression line) is: _____

Where the slope, _____ , represents:

And the y-intercept, _____ , represents:

The slope _____ has an interpretation, but the y-intercept will not have an interpretation if _____ is done.

To use the model to estimate, we simply evaluate the model at that point. This estimation found represents the _____ value of y for _____ x at that level.

236   Introduction to Statistics

A CI for β_1:

Interpretation:

A CI for $\mu_{y|x}$:

Interpretation:

A PI:

Interpretation:

Recall, the CI represents the _____ value of y for _____ x at that level, whereas a PI represents the value of y for _____ x at that level.

The correlation coefficient, denoted by the symbol _____, represents the strength of the _____ relationship between the variables.

Simple Linear Regression **237**

The values of the correlation coefficient are _____.

If the correlation coefficient is near zero that implies that there is

_____ relationship between the variables.

If the correlation coefficient is near positive or negative one, then that implies that there is a

_____. relationship between the variables.

The coefficient of determination, denoted by the symbol _____, represents the

strength of the _____ relationship between the variables.

The values of the coefficient of determination are _____.

There are two ways to interpret the coefficient of determination:

1. The percentage of the variation in _____ that can be explained by the _____
    relationship between x and y.

2. The percentage that we reduce our _____ using

    the linear regression equation to predict _____ versus using _____.

238   Introduction to Statistics

The assumptions of the regression model:

1.

2.

3.

4.

We verify the assumptions by doing the following three steps:

1.

2.

3.

Simple Linear Regression **239**

➡️ Chapter 9 Review Homework:

Consider the following scenario to answer the questions below:

Researchers would like to determine, for adults, if there is a linear relationship between number of hours of sleep obtained per night and weight (in pounds). If so, the researchers would like a model that will estimate the weight of adults given the number of hours of sleep obtained per night. To answer the questions, the following data was obtained:

Weight	Hours of Sleep	Weight	Hours of Sleep
143	5.25	109	8.75
135	6.55	131	7
140	4	109	9
122	6.3	122	7.75
121	7.15	112	8.75
122	7	133	7.2
130	5	115	8
138	4.35	112	9
146	3.5	120	8.5
121	8.11	106	9.25

1. What are the dependent and independent variables?

2. Is the relationship between the variables linear? If so, find the regression equation.

3. If appropriate, estimate the weight for someone who gets 6.5 hours of sleep per night. Interpret the estimate.

4. If appropriate, estimate the weight for someone who gets 1 hour of sleep per night. Interpret the estimate

5. Find and interpret a 95% confidence interval for how the average weight changes for a one hour increase in sleep per night.

6. Find and interpret a 95% confidence interval for the average weight given 6.5 hours of sleep per night is obtained.

7. Find and interpret a 95% prediction interval for the weight of a person who gets 6.5 hours of sleep per night.

8. Find and interpret the correlation coefficient.

9. Find and interpret the coefficient of determination.

Simple Linear Regression **241**

ANSWERS TO THE CHECKING YOUR UNDERSTANDING PROBLEMS

➡ Chapter 1:

Section 1:

Suppose Dr. Veen, a well-known oncologist, is interested in determining the average age of women when first diagnosed with breast cancer. Dr. Veen randomly samples 100 women that have been diagnosed with breast cancer and calculates their average age.

a. What is the parameter?

The average age of women when first diagnosed with breast cancer

b. What is the statistic?

The average age when first diagnosed with breast cancer for the 100 randomly sampled women.

Section 2:

1. For the following scenario, determine the type of method or sampling scheme used to obtain the data.

    a. John went to the food court of the mall and noted whether the patrons ate their purchased food at the food court or took their food to go.

    Since John just "noted", i.e., "observed", this is an Observational Study.

    b. Director Jones surveyed ten teachers from each school in Region 12 to determine how they feel about the new state requirements on teacher certification.

    The schools are the "stratas". Since Jones surveyed from each school, this is a Stratified Sample.

    c. Director Smith randomly selected 20 schools from his district and surveyed every teacher in the selected school to see if they support the new requirements on student grading policies.

    The schools are the "clusters". Since Smith surveyed every teacher at the randomly selected school, this is a Cluster Sample.

Answers to the Checking Your Understanding Problems **243**

Section 4:

Use the following scenario to answer the questions below:

Suppose the average age of a women diagnosed with breast cancer is 62 years old with standard deviation 2.4 years. Suppose Dr. Veen, a well-known oncologist, has treated 100 women in his career. His records show that the average age of his patients is 59 years old with standard deviation .1.8 years.

a. What is the value of $\bar{x}$?

$\bar{x}$ is the sample mean, thus $\bar{x} = 59$

b. What is the value of σ?

σ is the population standard deviation, thus $\sigma = 2.4$

c. Which measure describes the typical value of the population?

The typical value is a Measure of Central Tendency. Since for the population, and based on what we are given, the answer is the population mean, $\mu = 62$

d. Which measure describes how much variability there is about the sample mean?

Variability about the sample mean is the sample standard deviation or sample variance. We were given the sample standard deviation, $s = 1.8$

➡ Chapter 2:

Section 1:

Determine whether the following probability statements are valid. If invalid, state why.

a. $P(\mu > 1.2) = 0.0014$

Invalid. μ is the population mean and you cannot find probability about a parameter.

b. $P(X < 1) = -0.0023$

Invalid. Probability cannot be negative.

c. $P(X > 2) = 1.0011$

Invalid. Probability cannot be greater than 1.

Section 3:

1. Suppose the cumulative distribution function for the random variable X is as follows

X	0	0.5	1	1.5	2	4
$F(x)$	0.47	0.62	0.73	a	0.98	b

a. If the $P(X = 1.5) = 0.08$ what is the value of a?

$a = P(X \le 1.5) = P(X < 1.5) + P(X = 1.5) = P(X \le 1) + P(X = 1.5) = 0.73 + 0.08 = 0.81$

244 Introduction to Statistics

b. What is the value of b?

$b = P(X \le 4) = F(4) = 1$

c. $P(X = 2.5) = 0$ since $X = 2.5$ cannot occur.

We know it cannot occur since it is not listed in the table.

d. Find the probability that X is at least 1.5 hours in a given day.

$P(X \ge 1.5) = 1 - F(1) = 1 - 0.73 = 0.27$

e. $P(0 < X \le 2) = F(2) - F(0) = 0.98 - 0.47 = 0.51$

2. Suppose you play a dice game where if you roll a composite number, you win $5.00. If you roll a one, you win $10. If you roll a prime number, you lose $8. You must pay $2.00 to play. Let X = net winnings.

a. Find the PMF.

X	-10	3	8
f(x)	$^3/_6$	$^2/_6$	$^1/_6$

(Typically $^3/_6$ is reduced to ½ and $^2/_6$ is reduced to ⅓)

b. Find the expected value.

$E(X) = -10(½) + 3(⅓) + 8(⅙) = -^{16}/_6 = -2.67$

c. Interpret the expected value.

If we play this game over and over again, in the long run, we can expect to lose $2.67 per game.

Section 4:

Montana Table Company produces custom-made furniture for restaurants, diners and hotels. All of their furniture is handmade with it being known that 15 percent of the tables have some sort of flaw. Smitty's Steak House, a large restaurant, orders 15 new tables from Montana Table Company.

a. What is the probability that at least three will be flawed?

X = the number of tables flawed $\sim Bin(15, 0.15)$

(Use JMP to create the Binomial CDF table with $n = 15$ and $p = 0.15$)

$P(X \ge 3) = 1 - F(2) = 1 - 0.6042 = 0.3958$

b. How many tables ordered by Smitty's Steak House would you expect to have a flaw?

Since X has a binomial distribution, $E(X) = n \cdot p = 15 \cdot 0.15 = 2.25$

➡ Chapter 3:

Section 1:

1. For a continuous RV, $P(a < X < b) =$ <u>area under the PDF curve from a to b</u>.

2. Suppose you have the following PDF for the continuous RV, X. Find $P(X > 4)$

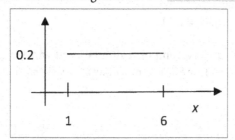

$P(X > 4) =$ area under the PDF curve from 4 to 6
$=$ area of the rectangle
$=$ base $\times$ height $= (6 - 4)(0.2) = 0.4$

Section 2:

Suppose you have the following CDF for the random variable, X.

$$F(x) = \begin{cases} 0 & X < 1 \\ \frac{3}{50}\left[\frac{x^3}{3} + x^2 - \frac{4}{3}\right] & 1 \leq X \leq 3 \\ 1 & X > 3 \end{cases}$$

a. Find $P(X \leq 1) = F(1) = 0$

b. Find $P(X > 2)$

$P(X > 2) = 1 - F(2) = 1 - \frac{3}{50}\left[\frac{2^3}{3} + 2^2 - \frac{4}{3}\right] = 1 - \frac{3}{50}\left[\frac{8 + 12 - 3}{3}\right]$

$= 1 - \frac{3}{50}\left[\frac{17}{3}\right] = 1 - \frac{17}{50} = 0.66$

Section 3:

Sue's Diner orders their chairs from Montana Table Company. When Sue calls Montana Table Company to place an order, she spends 5 to 30 minutes waiting on hold where the probability of waiting on hold is constant for all minutes between 5 and 30. Suppose Sue is interested in calculating the probabilities about the length of time she waits on hold.

a. Find the PDF for X.

$X =$ length of time on hold between 5 and 30 minutes $\sim Unif(5, 30)$

$f(x) = 1/(30-5) = 1/25$. So we have for the PDF, $f(x) = 1/25 \quad 5 < X < 30$.

b. Find the probability that Sue spends between 10 and 20 minutes on hold.

$P(10 < X < 20) =$ area of the rectangle $= b \cdot h = (20 - 10) \cdot (1/25) = 10/25 = 2/5$

Section 4:

1. Find $Z_{0.90}$

Find the z-score that has area to the right equal to 0.90. This corresponds to a z-score that has area in the left tail equal to 0.10. This is a special z-score. Thus, $Z_{0.90} = -1.282$

2. Find $Z_{0.30}$

Find the z-score that has area to the right equal to 0.30. This corresponds to the z-score that has area in the right tail equal to 0.30. This is not a special z-score so we have to use the $N(0,1)$ CDF table. Since the $N(0,1)$ CDF table gives area to the left, we know that the z-score that has area to the right equal to 0.30 has area to the left equal to 0.70. Thus we we look up AREA 0.70 in the $N(0,1)$ CDF table. Hence, $Z_{0.30} = 0.52$. (This is the z-score that has area to the left closest to 0.70)

3. The heights of cottonwood trees in Texas are normally distributed with a mean of 90 feet and standard deviation of 10 feet.

X = heights of cottonwood trees in Texas $\sim N(90, 10)$

    a. Find the probability that a randomly selected cottonwood tree is shorter than 75 feet.

$$P(X < 75) = P(Z < -1.5) = 0.0668$$

$$z = \frac{75 - 90}{10} = -1.5$$

    b. What is the probability that a randomly selected cottonwood tree is between 70 and 105 feet tall?

$$P(70 < X < 105) = P(-2 < X < 1.5) = 0.9332 - 0.0228 = 0.9104$$

$$z = \frac{70 - 90}{10} = -2 \text{ and } z = \frac{105 - 90}{10} = 1.5$$

    c. Find the probability that a randomly selected cottonwood tree is taller than 97 feet.

$$P(X > 97) = P(Z > 0.7) = 1 - 0.7580 = 0.2420$$

$$z = \frac{97 - 90}{10} = 0.7$$

    d. Find the height such that only 10% of trees are taller than that.

$$P(X > x) = 0.10$$

Find the specific x that has area to the right equal to 10%. Thus, we need to find the z-score that has area in the right tail equal to 10%. This is the special z-score = 1.282. Thus $X = 90 + 1.282(10) = 102.82$ ft

➡ Chapter 4:

Section 2:

1. Suppose it is known that adolescents play an online game for an average of 81 minutes per day with standard deviation of 22 minutes. If a random sample of 40 adolescents is polled, find the probability that the mean minutes spent playing an online game per day is less than 92 minutes.

Since $n = 40 \geq 30$, by CLT, $\overline{X} \sim N\left(81, \frac{22}{\sqrt{40}}\right)$

Answers to the Checking Your Understanding Problems **247**

$$P(\overline{X} < 92) = P(Z < 3.16) = 0.9992$$

$$Z = \frac{92 - 81}{22 / \sqrt{40}} = 3.16$$

2. Suppose the life spans of exercise bicycles are known to be normally distributed with a mean of 8.25 years and a standard deviation of 3.25 years.

    a. What is the probability that the mean life span in a group of 10 randomly selected exercise bicycles is between 7 and 11 years?

        Since we are told $X \sim N(8.25, 3.25)$, then by the Thm, $\overline{X} \sim N\left(81.25, \dfrac{3.25}{\sqrt{10}}\right)$

$$P(7 < \overline{X} < 11) = P(-1.22 < Z < 2.68) = 0.9963 - 0.1112 = 0.8851$$

$$Z = \frac{7 - 8.25}{3.25 / \sqrt{10}} = -1.22 \text{ and } Z = \frac{11 - 8.25}{3.25 / \sqrt{10}} = 2.68$$

    b. What is the probability that a randomly selected exercise bicycle has a life span of less than 13 years?

$$P(X < 13) = P(Z < 1.46) = 0.9279$$

$$Z = \frac{13 - 8.25}{3.25} = 1.46$$

Section 3:

1. Suppose that 35% of all high school students take shop class. If you sample 10 high school students, what is the probability that more than 30% of the students you sample take shop class?

    $P(\hat{p} > 0.30)$. But since $n \cdot p = 10 \cdot 0.35 = 3.5 < 5$, we cannot work as a $\hat{p}$ problem. We must solve it in terms of X where $X \sim Bin(10, 0.35)$ and $x = n \cdot \hat{p} = 10(0.30) = 3$. Thus, we must get the Binomial CDF from JMP.

    $P(\hat{p} > 0.30) = P(X > 3) = 1 - F(3) = 1 - 0.513827 = 0.4862$

2. Suppose that 30% of all retired people take an online class. If you sample 100 retired people, what is the probability that more than 35% of the them are taking an online class?

$$P(\hat{p} > 0.35) = P(Z > 1.09) = 1 - 0.8621 = 0.1379$$

$$Z = \frac{0.35 - 0.30}{\sqrt{\dfrac{0.30 \cdot 0.70}{100}}} = 1.09$$

➡ Chapter 5:

Section 1:

Researchers are interested in the proportion of students that are required to take an advanced level course in statistics prior to being admitted to nursing school. To determine, a random sample of 100 pre-nursing students is obtained and of those sampled 56 are required to take an advanced level course in statistics in order to gain admittance into nursing school.

a. What is the parameter of interest?

p = proportion of students required to take an advance level course in statistics prior to being admitted to nursing school.

b. What is the point estimate you would calculate to estimate the parameter? What is its value?

$\hat{p} = \dfrac{56}{100} = 0.56$

c. Is the point estimate an unbiased estimator of the parameter? Why or why not?

Yes, since $E[\hat{p}] = p$.

Section 2:

1. Suppose $E = 10$. Find the length of the confidence interval.

L = 2(10) = 20.

2. Suppose the length of the confidence interval is 9 and the point estimate is 12. Find the confidence interval.

Recall, E = L/2. Thus, E = 9/2 = 4.5

CI = (pt. est. − E, pt.est. + E) = (12 − 4.5, 12 + 4.5) = (7.5, 16.5)

Section 3:

Suppose the administrator in charge of student tickets is interested in estimating the proportion of students who left before half time at the last football game. This administrator randomly selects 60 students who attended the last game and asks if they left before half time. Of the 60 students, 32 said yes, they did.

a. Find a 95% confidence interval for the true proportion of students who left before half time at the last football game.

Since $n^*\hat{p} = 60*32/60 = 32 \geq 5$ and $n^*\hat{q} = 60*28/60 = 28 \geq 5$, the formula is valid.

95% CI for p: ${}^{32}\!/_{60} \pm 1.96 \cdot \sqrt{\dfrac{{}^{32}\!/_{60} \cdot {}^{28}\!/_{60}}{60}} = (0.4071, 0.6596)$

b. Can the administrator claim with 95% confidence that a majority of the students left before half time of the last game? Explain your answer.

No, the administrator cannot make that claim, since there are values contained within the CI that are less than 50%. (Recall, a majority means "more than half".)

Section 4:

Suppose it is believed that 36% of homes in the United States subscribe to some type of home viewing movie service. We would like to estimate the true proportion of homes in the US that subscribe to some type of home viewing movie service. What sample size should we use to ensure we estimate the truth within 3% with 95% confidence?

$n = \left(\dfrac{1.96}{0.03}\right)^{2} \cdot 0.36 \cdot 0.64 = 983.45$

Thus $n = 984$.

Answers to the Checking Your Understanding Problems **249**

Section 5:

1. Suppose it was known that people talked on their home phone for an average of 21 minutes per day. Researchers want to know if having such easy access to cell phones has changed the amount of time people talk on a phone. Thus, they would like to estimate with 95% confidence the average number of minutes that people talk on a phone (cell or land). Suppose a random sample of 40 people is polled, and the mean number of minutes talked on a land line or cell phone is 24 minutes with standard deviation 5 minutes.

   a. Estimate with 95% confidence the average number of minutes that people talk on a phone (cell or land)

   CI for μ: (we will assume the population is normally distributed. Since we do not have the data we cannot verify this assumption)

   $\bar{x} \pm t_{\alpha/2} \cdot s/\sqrt{n} = 24 \pm 2.023 \cdot 5/\sqrt{40} = (22.40, 25.60)$

   b. Can you be 95% confident that the average number of minutes people talk on a phone (cell or land) is more than 21 minutes? Explain your answer.

   Yes, since all the values in the CI are greater than 21.

2. Suppose a city architect is interested in the heights of the buildings in his town. He would like to know if the average height of the buildings in his town is different than 105 ft. The architect randomly selects 10 buildings and their heights, in feet, are listed below.

   106 109 107 108 112 113 104 105 115 114

   a. Find a 99% confidence interval for the average height of the buildings.

   Since we have the data, we will use JMP to find the CI.

   Since the data values are closely clustered to the line, we may assume that the population is normally distributed.

   A 99% CI is (105.25, 113.35)

   Confidence Intervals

Parameter	Estimate	Lower CI	Upper CI	1-Alpha
Mean	109.3	105.2453	113.3547	0.990
Std Dev	3.945462	2.437031	8.986234	0.990

   b. Can the city architect conclude with 99% confidence that the average height is different than 105 ft.

   Yes, since all the values within the CI are different than 105.

250   Introduction to Statistics

⮕ Chapter 6:

Section 2:

1. Suppose the observed value equals 2.34 and the rejection region is $Z_p \geq 1.645$. What is the decision of the test?

   Since 2.34 is greater than 1.645, the OV is the RR, thus we Reject Ho.

2. Suppose the p-value of the test equals 0.0215 and the level of significance equals 5%. What is the decision of the test?

   Since 2.15% < 5%, the p-value is less than α, thus we Reject Ho.

3. Suppose the following null and alternative hypothesis is being tested: Ho: $p = 0.5$ versus Ha: $p > 0.5$. Further suppose the decision of the test is Reject Ho. Which bound of the confidence interval should be constructed?

   Recall, the CI must agree with the Ha. $LB < p <$ UB Reading from the middle, we know p is greater than the lower bound, thus we just want to find the lower bound.

Section 3:

The Southern Online Viewing (SOW) Company claims that 60% of online video subscribers recommend its viewing platform over that of the Northern Online Viewing (NOW) Company's platform. To test this claim against the alternative that the actual proportion is less than 60%, a random sample of 50 online viewers was taken and 52% said they prefer SOW's platform over NOW's. At the 5% significance level, what conclusion can be reached?

1. Pop = online video subscribers

   POI = p = proportion of viewers that prefer SOW over NOW

2. α = 0.05

3. Ho: $p = 0.60$ versus   Ha: $p < 0.60$

4. Z_p-test $\sim N(0,1)$

5. $n \cdot p = 50(0.60) = 30 > 5$ and $n \cdot q = 50(0.40) = 20 > 5$

6. pt. est.: $\hat{p} = 0.52$. Thus our OV is $Z_p = \dfrac{0.52 - 0.60}{\sqrt{\dfrac{0.60 \cdot 0.40}{50}}} = -1.15$

7. RR: $Z_p \leq -1.645$ p-value: $P(\hat{p} \leq 0.52) = P(Z \leq -1.15) = 0.1251$

8. Fail to Reject Ho. At the 5% significance level there is not enough evidence to conclude that the proportion of customers that favor SOW over NOW is less than 60%.

Answers to the Checking Your Understanding Problems **251**

Section 4:

1. Suppose you have the following null and alternative hypothesis: Ho: It is too cold to ski

Ha: It is not too cold to ski

a. In the context of this problem, describe the power of the test.

Power = the probability of rejecting a false null hypothesis. So in the context of this problem, power is the probability of concluding that is it NOT too cold to ski when in reality it is NOT too cold to ski.

b. In the context of this problem, describe what α represents.

α = probability of making a type-I error, where a type-I error is rejecting a true null hypothesis. So in the context of this problem, α is the probability of concluding that it is NOT too cold to ski when in reality it IS too cold to ski.

c. In the context of this problem, describe what β represents.

β = probability of making a type-II error, where a type-II error is failing to reject a false null hypothesis. So in the context of this problem, β is the probability of concluding that it IS too cold to ski when in reality it is NOT too cold to ski.

2. To increase power, should n be increased or decreased?

Increased

3. To increase power, should α be increased or decreased?

Increased

4. Suppose the critical value of a two-tailed test equals $|2.064|$.

a. What is the rejection region?

$TS \leq -2.064$ or $TS \geq 2.064$

(where TS should be the symbol used to represent the test statistic used)

b. Suppose the observed value equals -3.45. What is the decision of the test?

The decision should be Reject Ho. (Since the OV is in the RR)

5. Suppose the p-value equals 0.0447 and the level of significance equals 5%. What is the decision of the test?

The decision should be Reject Ho. (Since the p-value is less than α)

Section 5:

1. The Southern Online Viewing (SOW) Company would like to see if they are increasing their average length of viewing per customer over last year. The average amount of SOW content watched by a subscriber per day last year was 1.8 hours. A random sample of 68 viewers found that the average length of viewership was 1.85 hours with a standard deviation of 0.45 hours. At the 5% significance level, is there significant evidence to conclude the claim from SOW? Assume all necessary assumptions are valid.

1. Pop = SOW customers   POI = μ = average length of viewing per customer

2. $\alpha = 0.05$

252   Introduction to Statistics

3. Ho: $\mu = 1.8$ versus Ha: $\mu > 1.8$
4. t-test $\sim t_{(67)}$
5. We have to assume the population is normally distributed.
6. pt. est.: $\bar{x} = 1.85$. Thus our OV is $t = \dfrac{1.85 - 1.8}{0.45/\sqrt{68}} = 0.92$
7. RR: $T \geq 1.668$ p-value: $P(\bar{x} \geq 1.85) = (T \geq 0.92) > 0.10$
8. Fail to Reject Ho. At the 5% significance level there is not enough evidence to conclude that the average length of viewing per customer is greater than 1.8 hours.

2. Suppose a researcher in public health is interested in determining if the average starting salary for those with a master's degree in public health is changing. The data below represents the starting salaries for 22 recent graduates that received a master's degree in public health. At the 1% significance level, can the researcher conclude that the average starting salary for those with a master's degree in public health is different from $55,000?

66.251 67.441 53.159 56.017 61.264 64.281 61.488 63.665 66.792 63.107 53.636
55.236 54.651 61.213 58.26   55.28   57.421 55.102 44.188 49.53   45.767 60.564

1. Pop = adults with a master's degree in public health
   POI = μ = average starting salary
2. $\alpha = 0.01$
3. Ho: $\mu = 55$ versus Ha: $\mu \neq 55$
4. t-test $\sim t_{(21)}$
5. Since the data values are closely clustered to the line, we may assume the population is normally distributed.
6. pt. est.: $\bar{x} = 57.9233$. Thus our OV is $t = 2.1360$
7. RR: $T \leq -2.831$ or $T \geq 2.831$ p-value: $2P(\bar{x} \geq 57.9233) = 0.0446$
8. Fail to Reject Ho. At the 1% significance level there is not enough evidence to conclude that the average starting salary for adults with a master's degree in public health is different than $55,000.

Distributions
Column 1

Test Mean

Hypothesized Value	55
Actual Estimate	57.9233
DF	21
Std Dev	6.41928

	t Test
Test Statistic	2.1360
Prob > |t|	0.0446*
Prob > t	0.0223*
Prob < t	0.9777

Section 6:

An educational researcher wants to determine if the percentage of high school students who like math is different from 50%. She takes a random sample of 85 high school students and finds that 32 of them like math.

a. What is the null and alternative hypothesis that corresponds to this problem?

Ho: $p = 0.50$ vs   Ha: $p \neq 0.50$

b. Find a 95% confidence interval for the proportion of high school students who like math.

Since $n \cdot p = 85(0.50) = n \cdot q = 42.5 \geq 5$, the formula is valid.

95% CI for p: $\dfrac{32}{85} \pm 1.96 \cdot \sqrt{\dfrac{{}^{32}\!/_{85} \cdot {}^{53}\!/_{85}}{85}} = (0.2735, 0.4795)$

c. Based on the confidence interval found in part b, what would be the decision of the hypothesis test for part a?

The decision is Reject Ho since the value 0.50 is not contained within the CI.

➡ Chapter 7:

Section 1:

Researchers are interested in comparing two different surgical methods for correcting cataracts. A random sample of 20 Americans that have cataracts in both eyes was obtained. Phacoemulsification surgery was performed on one eye and Extracapsular surgery was performed on the other eye of the randomly chosen subject. Researchers were interested in determining which method was most successful in correcting the cataract with minimal pain and side effects. Is this an example of independent or dependent samples?

Since the different surgeries were performed on the same person, one on each eye, this is an example of dependent samples.

Section 2: _

An article written in Healthy Choices magazine claimed that you would gain weight once you subscribe to an online movie repository site. Data analysts for the Southern Online Viewing (SOW) Company wanted to prove the article wrong, so they surveyed 20 new subscribers and measured their weight. After two years they followed up with them to measure their weight again. The data is given below. At the 1% significance level, do the analysts have evidence to support the article's claim?

Weight Before																			
128	112	124	116	119	127	115	126	121	120	119.5	128	115.5	121	119.5	117.5	120.5	118.5	111	122

Weight After																			
129.5	118	130.5	118	128	133	120.5	131	120	118	126	129.5	124	125	117	122	124.5	116.5	114.5	129

1. Pop = online video subscribers

   POI = μ_d where μ_1 = average weight before subscribing
   μ_2 = average weight after subscribing and $\mu_d = \mu_1 - \mu_2$

2. $\alpha = 0.01$

3. Ho: $\mu_d = 0$ versus   Ha: $\mu_d < 0$

4. t-test $\sim t_{(19)}$

5. Since the data values are closely clustered to the line we can assume that the differenced data is normally distributed.

6. pt. est.: $\bar{x}_d = -3.675$. Thus our OV is $t = -4.7019$

7. RR: $T \leq -2.539$ p-value: $P(\bar{x}_d \leq -3.675) < 0.0001$

8. Reject Ho. At the 1% significance level there is enough evidence to conclude that the average weight of new subscribers to an online movie site does increase once they join the site.

9. $\mu_d < -1.69$ I am 99% confident that the average weight gained for adults after joining an online movie subscription site is more than 1.69 pounds, on average. i.e., the average weight after joining an online movie site is more than 1.69 pounds greater than the average weight before joining an online movie site, on average.

Answers to the Checking Your Understanding Problems   **255**

Distributions

*Distributions
d in test scores*

Test Mean	
Hypothesized Value	0
Actual Estimate	-3.675
DF	19
Std Dev	3.49539
t Test	
Test Statistic	4.7019
Prob > |t|	0.0202*
Prob > t	0.9999
Prob < t	<.0001*

One-sided Confidence Interval				
Parameter	Estimate	Lower CI	Upper CI	1-Alpha
Mean	-3.675		-1.69016	0.990
Std Dev	3.495392		5.514837	0.990

Section 3:

Suppose you are interested in determining if men make more money than women, on average. To test, a random sample of ten men and ten women is obtained. At the 5% significance level, can you conclude that the average salary for men is higher than it is for women? The data given below represents the salaries in thousands.

Men	67.441	53.159	56.017	61.264	64.281	61.488	63.665	66.792	63.107	53.636
Women	54.651	61.213	58.26	55.28	57.421	55.102	44.188	49.53	45.767	60.564

1. Pop = adults

   POI = $\mu_2 - \mu_1$ where μ_1 = average salary for men and
   μ_2 = average salary for women

2. $\alpha = 0.05$

3. Ho: $\mu_2 = \mu_1$ versus Ha: $\mu_2 < \mu_1$

4. t_p-test ~ $t_{(18)}$

5. Verifying both populations are normally distributed: Since the data values are closely clustered to the line, we may assume both populations are normally distributed.

   Verifying equal population variances: Ho: $\sigma_1^2 = \sigma_2^2$ versus Ha: $\sigma_1^2 \neq \sigma_2^2$.
   Since the Brown-Forsythe p-value = 0.8149 > 0.10, we fail to reject Ho and may assume the population variances are equal.

6. $t_p = -2.78608$

7. RR: $T_p \leq -1.734$ p-value: $P((\bar{x}_2 - \bar{x}_1) \leq -6.887) = 0.0061$

8. Reject Ho At the 5% significance level, the evidence suggest that the average salary for men is greater than the average salary for women.

9. Since Ha: $\mu_2 < \mu_1$, we just want the upper bound. $-6.887 + 1.734(2.472) = -2.6006$
Thus, CI: $\mu_2 - \mu_1 < -2.6006$.
I am 95% confident that the average salary for men is greater than the average salary for women by an amount more than $2,600.60, on average.

Oneway Analysis of salaries By x

Oneway Anova

t Test

2-1

Assuming equal variances			
Difference	-6.887	t Ratio	-2.78608
Std Err Dif	2.472	DF	18
Upper CL Dif	-1.694	Prob > |t|	0.0122*
Lower CL Dif	-12.081	Prob > t	0.9939
Confidence	0.95	Prob < t	0.0061*

Means for Oneway Anova

Level	Number	Mean	Std Error	Lower 95%	Lower 95%
1	10	61.0850	1.7480	57.413	64.757
2	10	54.1976	1.7480	50.525	57.870

Tests that the Variances are Equal

Level	Count	Std Dev	MeanAbsDif to Mean	MeanAbsDif to Median
1	10	5.143670	4.088600	3.972200
2	10	5.886793	4.621560	4.350000

Test	F Ratio	DFNum	DFDen	p-Value
O'Brien[.5]	0.2862	1	18	0.5992
Brown-Forsythe	0.0564	1	18	0.8149
Levene	0.1510	1	18	0.7021
Bartlett	0.1548	1		0.6940
F Test 2-sided	1.3098	9	9	0.6942

Section 4:

Health researchers are interested in determining if more men smoke than women do. A random sample of 380 men included 97 who smoke, and a random sample of 345 women included 69 who smoke. Use a 0.1 significance level to test the claim that the proportion of men who smoke is greater than the proportion of women who smoke.

1. Pop = adults
   POI = $p_1 - p_2$ where p_1 = proportion of men that smoke
   $\qquad\qquad\qquad\quad p_2$ = proportion of women that smoke

2. $\alpha = 0.1$

3. Ho: $p_1 = p_2$ versus   Ha: $p_1 > p_2$

4. $n_1 \hat{p}_1 = 380\ ^{97}/_{380} = 97 \geq 5$ $n_2 \hat{p}_2 = 345\ ^{69}/_{345} = 69 \geq 5$

   $n_1 \hat{q}_1 = 380\ ^{283}/_{380} = 283 \geq 5$ $n_2 \hat{q}_2 = 345\ ^{276}/_{345} = 276 \geq 5$

5. Since JMP put p_2 first, we must think of our Ha in terms of p_2 first. So we have Ha: $p_2 < p_1$.
   So we have for the p-value: $P(\hat{p}_2 - \hat{p}_1 \leq 0) = 0.0387$

6. CI: Since two-tailed, the CI is testing p2 ≠ p1. The CI: (-0.1348, 0.0251).

7. Since the p-value is greater than 1% and the CI contains 0, our decision is Fail to reject Ho. Thus, at the 1% significance level, there is not enough evidence to suggest that the proportion of men that smoke is greater than the proportion of women that smoke.

Contingency Analysis of smoking status By sex

Two Sample Test for Proportions

Description	Proportion Difference	Lower 95%	Upper 95%
P(smoker|female)-P(smoker|male)	-0.05526	-0.13476	0.025134

Description	Proportion Difference
P(smoker|female)-P(smoker|male) ≥ 0	0.9613
P(smoker|female)-P(smoker|male) ≤ 0	0.0387*
P(smoker|female)-P(smoker|male) = 0	0.0774

258   Introduction to Statistics

Chapter 8:

Section 2:

Is the student grade classification at Baylor the same as it is for other universities? In the United States, of those attending a university, the percentage of freshmen students is 25%, the percentage of sophomore students is 16%, the percentage of college juniors is 18%, the percentage of seniors is 23%, and the percentage of graduate or "other" students is 18%. For the 2012/2013 school year, Baylor had 3739 freshmen, 2743 sophomores, 2849 juniors, 3399 seniors, and 2634 graduate or "other" students. At the 1% significance level, can you conclude that the proportions for Baylor differ for the US as a whole?

1. Pop = students
   POI = p_1 = proportion of freshmen, p_2 = proportion of sophomores,
   p_3 = proportion of juniors, p_4 = proportion of seniors,
   p_5 = proportion of graduate/ "other" students

2. $\alpha = 0.01$

3. Ho: $p_1 = 0.25, p_2 = 0.16, p_3 = 0.18, p_4 = 0.23, p_5 = 0.18$
   Ha: at least one differs from assumed

4. χ^2 - *Goodness of Fit* - test $\sim \chi^2_{(4)}$

5. $E_1 = 15,364(0.25) = 3,841 \geq 5$
   $E_2 = 15,364(0.16) = 2,458.24 \geq 5$
   $E_3 = 15,364(0.18) = 2,765.52 \geq 5$
   $E_4 = 15,364(0.23) = 3,533.72 \geq 5$
   $E_5 = 15,364(0.18) = 2,765.52 \geq 5$

6. $\chi^2 = 49.6057$

7. P-value: $P(\chi^2 \geq 49.6057) < 0.0001$

8. Reject Ho

9. Bonferroni CI: Let $\alpha^* = 0.01/5 = 0.002$. Have JMP create five 99.8% confidence intervals.
   p_1: (0.2328, 0.2542) Since CI contains 0.25, we cannot conclude p_1 is different than 25%
   p_2: (0.1692, 0.1883) Since CI does not contain 0.16, we can conclude p_2 is different than 16%
   p_3: (0.1759, 0.1953) Since CI contains 0.18, we cannot conclude p_3 is different than 18%
   p_4: (0.2111, 0.2318) Since CI contains 0.23, we cannot conclude p_4 is different than 23%
   p_5: (0.1622, 0.1810) Since CI contains 0.18, we cannot conclude p_5 is different than 18%

   At the 1% significance level, the only significant difference is with the proportion of sophomores. We are 99% confident that the proportion of sophomores is between 16.92% and 18.83%, on average.

Classification

Test Probabilities

Level	Estim Prob	Hypoth Prob
Freshmen	0.24336	0.25000
Junior	0.18543	0.18000
Others	0.17144	0.18000
Seniors	0.22123	0.23000
Sophomores	0.17853	0.16000

Test	ChiSquare	DF	Prob>Chisq
Likelihood Ratio	48.5684	4	<.0001*
Pearson	49.6057	4	<.0001*

Method:

Fix hypothesized values, rescale omitted

Confidence Intervals

Level	Count	Prob	Lower CI	Upper CI	1-Alpha
Freshmen	3739	0.24336	0.232824	0.254217	0.998
Junior	2849	0.18543	0.175941	0.195317	0.998
Others	2634	0.17144	0.162248	0.181039	0.998
Seniors	3399	0.22123	0.211058	0.231751	0.998
Sophomores	2743	0.17853	0.169187	0.188281	0.998
Total	15364				

Note: Computed using score confidence intervals.

Section 3:

A brokerage firm wants to see whether the type of account a customer has between the choices bronze, silver, gold, platinum, or diamond is related to who makes the trade, the customer him/herself or a professional trader hired by the person. A random sample of trades made for its customers over the past year was randomly chosen and it was found that for the trades made by a hired professional broker 55 were at the bronze level, 62 were silver, 75 were gold, 72 were platinum, and 83 were diamond. Of the trades made by the customer him/herself 75 were at the bronze level, 80 were silver, 82 were gold, 65 were platinum, and 67 were diamond. At the 2.5% significance level, what conclusion can be reached?

1. Pop = customers
   WTS: If the type of account between the choices bronze, silver, gold, platinum, or diamond is related to who made the trade; a professional or the customer him/herself.

2. $\alpha = 0.025$

3. Ho: independent   vs   Ha: dependent

260   Introduction to Statistics

4. $\chi^2 - Test\ of\ Independence - test \sim \chi^2_{(4)}$

5. From the output, $E_i \geq 5$ for each cell.

6. $\chi^2 = 7.066$

7. $P(\chi^2 \geq 7.066) = 0.1325$

8. Fail to Reject Ho. At the 2.5% significance level, there is not enough evidence to say that the type of trade made between the choices bronze, silver, gold, platinum, and diamond is related to who made the trade; the customer him/herself or a professional.

Contingency Analysis of type By made by Contingency Table

made by By type

Count Expected	bronze	diamond	gold	platinum	silver	Total
customer	75 66.9972	67 77.3045	82 80.912	65 70.6047	80 73.1816	369
professional	55 63.0028	83 72.6955	75 76.088	72 66.3953	62 68.8184	347
Total	130	150	157	137	142	716

Tests

N	DF	-LogLike	RSquare (U)
716	4	3.5404383	0.0031

Tests	ChiSquare	Prob>ChiSq
Likelihood Ratio	7.081	0.1317*
Pearson	7.066	0.1325*

Section 4:

Campus Life administrators want to determine if students are similar with respect to if and where they work. A random sample of freshman found that 320 do not work, 260 work as a student worker for the University, and 450 work off-campus. A random sample of sophomores found that 280 do not work, 220 work as a student worker for the University, and 350 work off-campus. A random sample of juniors found that 250 do not work, 315 work as a student worker for the University, and 380 work off-campus. A random sample of seniors found that 280 do not work, 220 work as a student worker for the University, and 350 work off-campus. At the 5% significance level what conclusion can be reached?

1. Pop = students
   WTS: If the students based on classification are similar with respect to work status.

2. $\alpha = 0.05$

3. Ho: homogeneous   vs   Ha: heterogeneous

4. $\chi^2 - Test\ of\ Homogeneity - test \sim \chi^2_{(6)}$

5. From the output, $E_i \geq 5$ for each cell.

Answers to the Checking Your Understanding Problems **261**

6. $\chi^2 = 25.082$

7. $P(\chi^2 \geq 25.082) = 0.0003$

8. Reject Ho. At the 5% significance level, the evidence suggests that freshmen, sophomores, juniors, and seniors are not similar with respect to work status.

Contingency Analysis of work status By classification

Contingency Table

classification By work status

Count Expected	no	off campus	on campus	Total
Fr	320 316.707	450 428.816	260 284.476	1030
Ju	250 290.571	380 393.429	315 261	945
Se	280 261.361	350 353.878	220 234.762	850
So	280 261.361	350 353.878	220 234.762	850
Total	1130	1530	1015	3675

Tests

N	DF	-LogLike	RSquare (U)
3675	6	12.348754	0.0031

Tests	ChiSquare	Prob>ChiSq
Likelihood Ratio	24.698	0.0004*
Pearson	25.082	0.0003*

➡ Chapter 9:

Section 3:

A pharmaceutical company has developed a new drug for patients with Wolfe-Parkinson-White syndrome designed to control paroxysmal supraventricular tachycardia (PSVT). Patients with PSVT have a heart rate of more than 100 beats per minute. (BPM) Suppose you are interested in modeling the relationship between the decrease in pulse rate in BPM and dosage in cc. Different doses were administered to 16 randomly selected patients and 20 minutes later the decrease in each patient's pulse rate was recorded. The data collected is given below.

Dosage	2	4	1.5	1	3	3.5	2.5	3	3.5	1.5	2	2	3.75	3.5	1.75	1.25
Drop in HR	14	24	11	6	18	22	17	15	22	9	16	14	23	25	12	12

a. Which variable is the independent variable?   Dosage

b. Which variable is the dependent variable?   Drop in heart rate

262   Introduction to Statistics

c. Is a linear model appropriate? Explain your answer.

Yes, since the data values are forming a linear shape.

d. If a linear model is appropriate, find the line of best fit.

$\hat{y} = 2.5983762 + 5.4949932x$

e. If appropriate, interpret the coefficients of the model.

$b_0 = 2.5984$ The y-intercept does not have any interpretation since we did not consider values near zero.

$b_1 = 5.4950$ If we increase the dosage by one cc, the average drop in heart rate will increase by 5.4950 beats per minute.

f. If appropriate, estimate the drop in pulse rate for a dosage level of 2.75 cc. Interpret the estimate found.

$\hat{y} = 2.5983762 + 5.4949932(2.75) = 17.71$

We estimate that anyone who receives a dosage of 2.75 cc will have an average drop in pulse rate of 17.71 BPM.

g. If appropriate, estimate the drop in pulse rate for a dosage level of 5.5 cc.

Since we did not consider values of X that big, it would be extrapolating too far beyond the range of X values considered and thus would not be appropriate to use the model to estimate for X = 5.5.

Bivariate Fit of y By x

Summary of Fit	
RSquare	0.898601
RSquare Adj	0.891359
Root Mean Square Error	1.87423
Mean of Response	16.25
Observations (or Sum Wgts)	16

Linear Fit

$y = 2.5983762 + 5.4949932*x$

Parameter Estimates				
Term	Estimate	Std Error	t Ratio	Prob>|t|
Intercept	2.5983762	1.312124	1.98	0.0677
x	5.4949932	0.493328	11.14	<.0001*

Section 4:

(Using the same information from Section 3 problem)

a. Find and interpret a 99% confidence interval for the slope of the regression line.

99% CI for β_1: $5.4950 \pm 2.977(0.4933) = (4.03, 6.96)$.

We are 99% confident that if we increase the dosage by one cc, the average drop in pulse rate will increase between 4.03 and 6.96 BPM, on average.

Term	Estimate	Std Error	t Ratio	Prob>|t|
Intercept	2.5983762	1.312124	1.98	0.0677
x	5.4949932	0.493328	11.14	<.0001*

b. Find and interpret a 99% confidence interval for the average drop in pulse rate for a dosage level of 2.75 cc.

99% CI for $\mu_{Y|x}$: $(16.26, 19.16)$

We are 99% confident that anyone who receives a dosage of 2.75 ccs will have an average drop in pulse rate between 16.26 and 19.16 BPM, on average.

(Use JMP to get CI. Make sure and define $\alpha = 0.01$)

1.25	12	7.179847771	11.754387682	3.4371867169	15.497048736
2.75	.	16.261264998	19.157950158	11.94539381	23.473821346

c. Find and interpret a 99% prediction interval for a dosage level of 2.75 cc.

PI: $(11.95, 23.47)$

We are 99% confident that this specific person who received a dosage of 2.75 cc will have a drop in pulse rate between 11.95 and 23.47 BPM, on average.

1.25	12	7.179847771	11.754387682	3.4371867169	15.497048736
2.75	.	16.261264998	19.157950158	11.94539381	23.473821346

(Use JMP to get PI. Make sure and define $\alpha = 0.01$)

Section 5:

(Using the same information from Section 3 problem)

a. Find the correlation coefficient. Is there a strong linear relationship? Is it a direct or an indirect?

From JMP we get $r^2 = 0.898601$. *Thus,* $r = \sqrt{0.898601} = 0.9479$.

(r is positive, since b_1 is positive.) Since r is close to 1, there is a strong direct linear relationship between dosage level and drop in pulse rate.

264 Introduction to Statistics

Summary of Fit	
RSquare	0.898601
RSquare Adj	0.891359
Root Mean Square Error	1.87423
Mean of Response	16.25
Observations (or Sum Wgts)	16

b. Find the coefficient of determination.

From JMP we get $r^2 = 0.8986$.

c. Interpret the coefficient of determination using the variance interpretation.

Approximately 89.86% of the variation in drop in pulse rate can be explained by the linear relationship between dosage level and drop in pulse rate.

d. Interpret the coefficient of determination using the error interpretation.

We reduce our prediction error by approximately 89.86% using the linear regression line to estimate drop in pulse rate versus just using the average pulse rate to estimate drop in pulse rate.

Section 6:

(Using the same information from Section 3 problem)

Perform residual analysis to verify the assumptions of the model.

Residual Analysis consists of three steps:

1. Create Residual Plot:

Diagnostics Plots
Residual by Predicted Plot

Since the residuals are randomly distributed around zero, i.e., there is no pattern, we can assume the following to be valid.

1. The relationship between dosage level and average drop in pulse rate is linear.

2. The errors are independent.

3. The errors have constant variance and zero mean.

2. QQ-plot of the residuals.

   Residual Normal Quantile Plot

   Since the residuals are closely clustered to the line, we may assume that they are normally distributed.

3. Check for Outliers.

   $\pm 3(S) = \pm 3(\text{Root Mean Square Error}) = \pm 3(1.87432) = \pm 5.62$

   The residuals found from JMP:

0.4116373478	-0.57834912	0.1591339648	-2.093369418
-1.083355886	0.1691474966	0.6641407307	-4.083355886
0.1691474966	-1.840866035	2.4116373478	0.4116373478
-0.204600812	3.1691474966	-0.214614344	2.5328822733

   Since none of the residuals are larger in magnitude to 5.62, we can conclude that there are not any outliers.

FORMULA/TABLE CARD FOR WEISS'S *INTRODUCTORY STATISTICS, EIGHTH EDITION*

Larry R. Griffey

TABLE II Areas under the standard normal curve

0.09	0.08	0.07	0.06	0.05	0.04	0.03	0.02	0.01	0.00	z
									0.0000*	−3.9
0.0001	0.0001	0.0001	0.0001	0.0001	0.0001	0.0001	0.0001	0.0001	0.0001	−3.8
0.0001	0.0001	0.0001	0.0001	0.0001	0.0001	0.0001	0.0001	0.0001	0.0001	−3.7
0.0001	0.0001	0.0001	0.0001	0.0001	0.0001	0.0001	0.0001	0.0002	0.0002	−3.6
0.0002	0.0002	0.0002	0.0002	0.0002	0.0002	0.0002	0.0002	0.0002	0.0002	−3.5
0.0002	0.0003	0.0003	0.0003	0.0003	0.0003	0.0003	0.0003	0.0003	0.0003	−3.4
0.0003	0.0004	0.0004	0.0004	0.0004	0.0004	0.0004	0.0005	0.0005	0.0005	−3.3
0.0005	0.0005	0.0005	0.0006	0.0006	0.0006	0.0006	0.0006	0.0007	0.0007	−3.2
0.0007	0.0007	0.0008	0.0008	0.0008	0.0008	0.0009	0.0009	0.0009	0.0010	−3.1
0.0010	0.0010	0.0011	0.0011	0.0011	0.0012	0.0012	0.0013	0.0013	0.0013	−3.0
0.0014	0.0014	0.0015	0.0015	0.0016	0.0016	0.0017	0.0018	0.0018	0.0019	−2.9
0.0019	0.0020	0.0021	0.0021	0.0022	0.0023	0.0023	0.0024	0.0025	0.0026	−2.8
0.0026	0.0027	0.0028	0.0029	0.0030	0.0031	0.0032	0.0033	0.0034	0.0035	−2.7
0.0036	0.0037	0.0038	0.0039	0.0040	0.0041	0.0043	0.0044	0.0045	0.0047	−2.6
0.0048	0.0049	0.0051	0.0052	0.0054	0.0055	0.0057	0.0059	0.0060	0.0062	−2.5
0.0064	0.0066	0.0068	0.0069	0.0071	0.0073	0.0075	0.0078	0.0080	0.0082	−2.4
0.0084	0.0087	0.0089	0.0091	0.0094	0.0096	0.0099	0.0102	0.0104	0.0107	−2.3
0.0110	0.0113	0.0116	0.0119	0.0122	0.0125	0.0129	0.0132	0.0136	0.0139	−2.2
0.0143	0.0146	0.0150	0.0154	0.0158	0.0162	0.0166	0.0170	0.0174	0.0179	−2.1
0.0183	0.0188	0.0192	0.0197	0.0202	0.0207	0.0212	0.0217	0.0222	0.0228	−2.0
0.0233	0.0239	0.0244	0.0250	0.0256	0.0262	0.0268	0.0274	0.0281	0.0287	−1.9
0.0294	0.0301	0.0307	0.0314	0.0322	0.0329	0.0336	0.0344	0.0351	0.0359	−1.8
0.0367	0.0375	0.0384	0.0392	0.0401	0.0409	0.0418	0.0427	0.0436	0.0446	−1.7
0.0455	0.0465	0.0475	0.0485	0.0495	0.0505	0.0516	0.0526	0.0537	0.0548	−1.6
0.0559	0.0571	0.0582	0.0594	0.0606	0.0618	0.0630	0.0643	0.0655	0.0668	−1.5
0.0681	0.0694	0.0708	0.0721	0.0735	0.0749	0.0764	0.0778	0.0793	0.0808	−1.4
0.0823	0.0838	0.0853	0.0869	0.0885	0.0901	0.0918	0.0934	0.0951	0.0968	−1.3
0.0985	0.1003	0.1020	0.1038	0.1056	0.1075	0.1093	0.1112	0.1131	0.1151	−1.2
0.1170	0.1190	0.1210	0.1230	0.1251	0.1271	0.1292	0.1314	0.1335	0.1357	−1.1
0.1379	0.1401	0.1423	0.1446	0.1469	0.1492	0.1515	0.1539	0.1562	0.1587	−1.0
0.1611	0.1635	0.1660	0.1685	0.1711	0.1736	0.1762	0.1788	0.1814	0.1841	−0.9
0.1867	0.1894	0.1922	0.1949	0.1977	0.2005	0.2033	0.2061	0.2090	0.2119	−0.8
0.2148	0.2177	0.2206	0.2236	0.2266	0.2296	0.2327	0.2358	0.2389	0.2420	−0.7
0.2451	0.2483	0.2514	0.2546	0.2578	0.2611	0.2643	0.2676	0.2709	0.2743	−0.6
0.2776	0.2810	0.2843	0.2877	0.2912	0.2946	0.2981	0.3015	0.3050	0.3085	−0.5
0.3121	0.3156	0.3192	0.3228	0.3264	0.3300	0.3336	0.3372	0.3409	0.3446	−0.4
0.3483	0.3520	0.3557	0.3594	0.3632	0.3669	0.3707	0.3745	0.3783	0.3821	−0.3
0.3859	0.3897	0.3936	0.3974	0.4013	0.4052	0.4090	0.4129	0.4168	0.4207	−0.2
0.4247	0.4286	0.4325	0.4364	0.4404	0.4443	0.4483	0.4522	0.4562	0.4602	−0.1
0.4641	0.4681	0.4721	0.4761	0.4801	0.4840	0.4880	0.4920	0.4960	0.5000	−0.0

* For $z \leq -3.90$, the areas are 0.0000 to four decimal places.

FORMULA/TABLE CARD FOR WEISS'S *INTRODUCTORY STATISTICS, EIGHTH EDITION*
Larry R. Griffey

TABLE II (cont.) Areas under the standard normal curve

z	0.00	0.01	0.02	0.03	0.04	0.05	0.06	0.07	0.08	0.09
0.0	0.5000	0.5040	0.5080	0.5120	0.5160	0.5199	0.5239	0.5279	0.5319	0.5359
0.1	0.5398	0.5438	0.5478	0.5517	0.5557	0.5596	0.5636	0.5675	0.5714	0.5753
0.2	0.5793	0.5832	0.5871	0.5910	0.5948	0.5987	0.6026	0.6064	0.6103	0.6141
0.3	0.6179	0.6217	0.6255	0.6293	0.6331	0.6368	0.6406	0.6443	0.6480	0.6517
0.4	0.6554	0.6591	0.6628	0.6664	0.6700	0.6736	0.6772	0.6808	0.6844	0.6879
0.5	0.6915	0.6950	0.6985	0.7019	0.7054	0.7088	0.7123	0.7157	0.7190	0.7224
0.6	0.7257	0.7291	0.7324	0.7357	0.7389	0.7422	0.7454	0.7486	0.7517	0.7549
0.7	0.7580	0.7611	0.7642	0.7673	0.7704	0.7734	0.7764	0.7794	0.7823	0.7852
0.8	0.7881	0.7910	0.7939	0.7967	0.7995	0.8023	0.8051	0.8078	0.8106	0.8133
0.9	0.8159	0.8186	0.8212	0.8238	0.8264	0.8289	0.8315	0.8340	0.8365	0.8389
1.0	0.8413	0.8438	0.8461	0.8485	0.8508	0.8531	0.8554	0.8577	0.8599	0.8621
1.1	0.8643	0.8665	0.8686	0.8708	0.8729	0.8749	0.8770	0.8790	0.8810	0.8830
1.2	0.8849	0.8869	0.8888	0.8907	0.8925	0.8944	0.8962	0.8980	0.8997	0.9015
1.3	0.9032	0.9049	0.9066	0.9082	0.9099	0.9115	0.9131	0.9147	0.9162	0.9177
1.4	0.9192	0.9207	0.9222	0.9236	0.9251	0.9265	0.9279	0.9292	0.9306	0.9319
1.5	0.9332	0.9345	0.9357	0.9370	0.9382	0.9394	0.9406	0.9418	0.9429	0.9441
1.6	0.9452	0.9463	0.9474	0.9484	0.9495	0.9505	0.9515	0.9525	0.9535	0.9545
1.7	0.9554	0.9564	0.9573	0.9582	0.9591	0.9599	0.9608	0.9616	0.9625	0.9633
1.8	0.9641	0.9649	0.9656	0.9664	0.9671	0.9678	0.9686	0.9693	0.9699	0.9706
1.9	0.9713	0.9719	0.9726	0.9732	0.9738	0.9744	0.9750	0.9756	0.9761	0.9767
2.0	0.9772	0.9778	0.9783	0.9788	0.9793	0.9798	0.9803	0.9808	0.9812	0.9817
2.1	0.9821	0.9826	0.9830	0.9834	0.9838	0.9842	0.9846	0.9850	0.9854	0.9857
2.2	0.9861	0.9864	0.9868	0.9871	0.9875	0.9878	0.9881	0.9884	0.9887	0.9890
2.3	0.9893	0.9896	0.9898	0.9901	0.9904	0.9906	0.9909	0.9911	0.9913	0.9916
2.4	0.9918	0.9920	0.9922	0.9925	0.9927	0.9929	0.9931	0.9932	0.9934	0.9936
2.5	0.9938	0.9940	0.9941	0.9943	0.9945	0.9946	0.9948	0.9949	0.9951	0.9952
2.6	0.9953	0.9955	0.9956	0.9957	0.9959	0.9960	0.9961	0.9962	0.9963	0.9964
2.7	0.9965	0.9966	0.9967	0.9968	0.9969	0.9970	0.9971	0.9972	0.9973	0.9974
2.8	0.9974	0.9975	0.9976	0.9977	0.9977	0.9978	0.9979	0.9979	0.9980	0.9981
2.9	0.9981	0.9982	0.9982	0.9983	0.9984	0.9984	0.9985	0.9985	0.9986	0.9986
3.0	0.9987	0.9987	0.9987	0.9988	0.9988	0.9989	0.9989	0.9989	0.9990	0.9990
3.1	0.9990	0.9991	0.9991	0.9991	0.9992	0.9992	0.9992	0.9992	0.9993	0.9993
3.2	0.9993	0.9993	0.9994	0.9994	0.9994	0.9994	0.9994	0.9995	0.9995	0.9995
3.3	0.9995	0.9995	0.9995	0.9996	0.9996	0.9996	0.9996	0.9996	0.9996	0.9997
3.4	0.9997	0.9997	0.9997	0.9997	0.9997	0.9997	0.9997	0.9997	0.9997	0.9998
3.5	0.9998	0.9998	0.9998	0.9998	0.9998	0.9998	0.9998	0.9998	0.9998	0.9998
3.6	0.9998	0.9998	0.9999	0.9999	0.9999	0.9999	0.9999	0.9999	0.9999	0.9999
3.7	0.9999	0.9999	0.9999	0.9999	0.9999	0.9999	0.9999	0.9999	0.9999	0.9999
3.8	0.9999	0.9999	0.9999	0.9999	0.9999	0.9999	0.9999	0.9999	0.9999	0.9999
3.9	1.0000*									

* For $z \geq 3.90$, the areas are 1.0000 to four decimal places.

FORMULA/TABLE CARD FOR WEISS'S *INTRODUCTORY STATISTICS, EIGHTH EDITION*

Larry R. Griffey

TABLE IV Values of t_α

df	$t_{0.10}$	$t_{0.05}$	$t_{0.025}$	$t_{0.01}$	$t_{0.005}$	df
1	3.078	6.314	12.706	31.821	63.657	1
2	1.886	2.920	4.303	6.965	9.925	2
3	1.638	2.353	3.182	4.541	5.841	3
4	1.533	2.132	2.776	3.747	4.604	4
5	1.476	2.015	2.571	3.365	4.032	5
6	1.440	1.943	2.447	3.143	3.707	6
7	1.415	1.895	2.365	2.998	3.499	7
8	1.397	1.860	2.306	2.896	3.355	8
9	1.383	1.833	2.262	2.821	3.250	9
10	1.372	1.812	2.228	2.764	3.169	10
11	1.363	1.796	2.201	2.718	3.106	11
12	1.356	1.782	2.179	2.681	3.055	12
13	1.350	1.771	2.160	2.650	3.012	13
14	1.345	1.761	2.145	2.624	2.977	14
15	1.341	1.753	2.131	2.602	2.947	15
16	1.337	1.746	2.120	2.583	2.921	16
17	1.333	1.740	2.110	2.567	2.898	17
18	1.330	1.734	2.101	2.552	2.878	18
19	1.328	1.729	2.093	2.539	2.861	19
20	1.325	1.725	2.086	2.528	2.845	20
21	1.323	1.721	2.080	2.518	2.831	21
22	1.321	1.717	2.074	2.508	2.819	22
23	1.319	1.714	2.069	2500	2.807	23
24	1.318	1.711	2.064	2.492	2.797	24
25	1.316	1.708	2.060	2.485	2.787	25
26	1.315	1.706	2.056	2.479	2.779	26
27	1.314	1.703	2.052	2.473	2.771	27
28	1.313	1.701	2.048	2.467	2.763	2S
29	1.311	1.699	2.045	2.462	2.756	29
30	1.310	1.697	2.042	2.457	2.750	30
31	1.309	1.696	2.040	2.453	2.744	31
32	1.309	1.694	2.037	2.449	2.738	32
33	1.308	1.692	2.035	2.445	2.733	33
34	1.307	1.691	2.032	2.441	2.728	34
35	1.306	1.690	2.030	2.438	2.724	35
36	1.306	1.688	2.028	2.434	2.719	36
37	1.305	1.687	2.026	2.431	2.715	37
38	1.304	1.686	2.024	2.429	2.712	38
39	1.304	1.685	2.023	2.426	2.708	39
40	1.303	1.684	2.021	2.423	2.704	40
41	1.303	1.683	2.020	2.421	2.701	41
42	1.302	1.682	2.018	2.418	2.698	42
43	1.302	1.681	2.017	2.416	2.695	43
44	1.301	1.680	2.015	2.414	2.692	44
45	1.301	1.679	2.014	2.412	2.690	45
46	1.300	1.679	2.013	2.410	2.687	46
47	1.300	1.678	2.012	2.408	2.685	47
48	1.299	1.677	2.011	20407	2.682	48
49	1.299	1.677	2.010	2.405	2.680	49

TABLE IV (cont.) Values of t_α

df	$t_{0.10}$	$t_{0.05}$	$t_{0.025}$	$t_{0.01}$	$t_{0.005}$	df
50	1.299	1.676	2.009	2.403	2.678	50
51	1.298	1.675	2.008	2.402	2.676	51
52	1.298	1.675	2.007	2.400	2.674	52
53	1.298	1.674	2.006	2.399	2.672	53
54	1.297	1.674	2.005	2.397	2.670	54
55	1.297	1.673	2.004	2.396	2.668	55
56	1.297	1.673	2.003	2.395	2.667	56
57	1.297	1.672	2.002	2.394	2.665	57
58	1.296	1.672	2.002	2.392	2.663	58
59	1.296	1.671	2.001	2.391	2.662	59
60	1.296	1.671	2.000	2.390	2.660	60
61	1.296	1.670	2.000	2.389	2.659	61
62	1.295	1.670	1.999	2.388	2.657	62
63	1.295	1.669	1.998	2.387	2.656	63
64	1.295	1.669	1.998	2.386	2.655	64
65	1.295	1.669	1.997	2.385	2.654	65
66	1.295	1.668	1.997	2.384	2.652	66
67	1.294	1.668	1.996	2.383	2.651	67
68	1.294	1.668	1.995	2.382	2.650	68
69	1.294	1.667	1.995	2.382	2.649	69
70	1.294	1.667	1.994	2.381	2.648	70
71	1.294	1.667	1.994	2.380	2.647	71
72	1.293	1.666	1.993	2.379	2.646	72
73	1.293	1.666	1.993	2.379	2.645	73
74	1.293	1.666	1.993	2.378	2.644	74
75	1.293	1.665	1.992	2.377	2.643	75
80	1.292	1.664	1.990	2.374	2.639	80
85	1.292	1.663	1.988	2.371	2.635	85
90	1.291	1.662	1.987	2.368	2.632	90
95	1.291	1.661	1.985	2.366	2.629	95
100	1.290	1.660	1.984	2.364	2.626	100
200	1.286	1.653	1.972	2.345	2.601	200
300	1.284	1.650	1.968	2.339	2.592	300
400	1.284	1.649	1.966	2.336	2.588	400
500	1.283	1.648	1.965	2.334	2.586	500
600	1.283	1.647	1.964	2.333	2.584	600
700	1.283	1.647	1.963	2.332	2.583	700
800	1.283	1.647	1.963	2.331	2.582	800
900	1.282	1.647	1.963	2.330	2.581	900
1000	1.282	1.646	1.962	2.330	2.581	1000
2000	1.282	1.646	1.961	2.328	2.578	2000

1.282	1.645	1.960	2.326	2.576
$z_{0.10}$	$z_{0.05}$	$z_{0.025}$	$z_{0.01}$	$z_{0.005}$

FORMULA/TABLE CARD FOR WEISS'S *INTRODUCTORY STATISTICS, EIGHTH EDITION*
Larry R. Griffey

TABLE VII   Values of χ^2_α

$\chi^2_{0.10}$	$\chi^2_{0.05}$	$\chi^2_{0.025}$	$\chi^2_{0.01}$	$\chi^2_{0.005}$	df
2.706	3.841	5.024	6.635	7.879	1
4.605	5.991	7.378	9.210	10.597	2
6.251	7.815	9.348	11.345	12.838	3
7.779	9.488	11.143	13.277	14.860	4
9.236	11.070	12.833	15.086	16.750	5
10.645	12.592	14.449	16.812	18.548	6
12.011	14.061	16.013	18.475	20.278	7
13.362	15.507	17.535	20.090	21.955	8
14.684	16.919	19.023	21.666	23.589	9
15.987	18.307	20.483	23.209	25.188	10
17.275	19.675	21.920	24.725	26.757	11
18.549	21.026	23.337	26.217	28.300	12
19.812	22.362	24.736	27.688	29.819	13
21.064	23.685	26.119	29.141	31.319	14
22.307	24.996	21.488	30.578	32.801	15
23.542	26.296	28.845	32.000	34.267	16
24.769	27.587	30.191	33.409	35.718	17
25.989	28.869	31.526	34.805	37.156	18
27.204	30.143	32.852	36.191	38.582	19
28.412	31.410	34.170	37.566	39.997	20
29.615	32.611	35.419	38.932	41.401	21
30.813	33.924	36.781	40.290	42.796	22
32.007	35.172	38.076	41.638	44.181	23
33.196	36.415	39.364	42.980	45.559	24
34.382	37.653	40.647	44.314	46.928	25
35.563	38.885	41.923	45.642	48.290	26
36.741	40.113	43.195	46.963	49.645	27
37.916	41337	44.461	48.278	50.994	28
39.087	42.557	45.722	49.588	52.336	29
40.256	43.773	46.979	50.892	53.672	30
51.805	55.759	59.344	63.691	66.167	40
63.161	61.505	71.420	76.154	79.490	50
74.397	19.082	83.298	88.381	91.955	60
85.527	90.531	95.023	100.424	104.213	70
96.578	101.879	106.628	112.328	116.320	80
107.565	113.145	118.135	124.115	128.296	90
118.499	124.343	129.563	135.811	140.177	100